Sustainable Farm Forestry in the Tropics

NEW HORIZONS IN ENVIRONMENTAL ECONOMICS

General Editors: Wallace E. Oates, *Professor of Economics, University of Maryland, USA* and Henk Folmer, *Professor of General Economics, Wageningen University and Professor of Environmental Economics, Tilburg University, The Netherlands*

This important series is designed to make a significant contribution to the development of the principles and practices of environmental economics. It includes both theoretical and empirical work. International in scope, it addresses issues of current and future concern in both East and West and in developed and developing countries.

The main purpose of the series is to create a forum for the publication of high quality work and to show how economic analysis can make a contribution to understanding and resolving the environmental problems confronting the world in the twenty-first century.

Recent titles in the series include:

Sustainable Small-scale Forestry
Socio-economic Analysis and Policy
Edited by S.R. Harrison, J.L. Herbohn and K.F. Herbohn

Environmental Economics and Public Policy
Selected Papers of Robert N. Stavins, 1988-1999
Robert N. Stavins

International Environmental Externalities and the Double Dividend
Sebastian Killinger

Global Emissions Trading
Key Issues for Industrialized Countries
Edited by Suzi Kerr

The Choice Modelling Approach to Environmental Valuation
Edited by Jeff Bennett and Russell Blamey

Uncertainty and the Environment
Implications for Decision Making and Environmental Policy
Richard A. Young

Global Warming and the American Economy
A Regional Assessment of Climate Change Impacts
Edited by Robert Mendelsohn

The International Yearbook of Environmental and Resource Economics 2001/2002
A Survey of Current Issues
Edited by Henk Folmer and Tom Tietenberg

Sustainable Farm Forestry in the Tropics
Social and Economic Analysis and Policy
Edited by S.R. Harrison and J.L. Herbohn

The Economic Value of Water Quality
Edited by John C. Bergstrom, Kevin J. Boyle and Gregory L. Poe

Implementing European Environmental Policy
The Impacts of Directives in the Member States
Edited by Matthieu Glachant

Sustainable Farm Forestry in the Tropics:

Social and Economic Analysis and Policy

Edited by

S.R. Harrison

School of Economics, The University of Queensland, Brisbane, Australia

J.L. Herbohn

School of Natural and Rural Systems Management, The University of Queensland, Gatton Campus, Australia

NEW HORIZONS IN ENVIRONMENTAL ECONOMICS

Edward Elgar

Cheltenham, UK • Northampton, MA, USA

Published by
Edward Elgar Publishing Limited
Glensanda House
Montpellier Parade
Cheltenham
Glos GL50 1UA
UK

Edward Elgar Publishing, Inc.
136 West Street
Suite 202
Northampton
Massachusetts 01060
USA

A catalogue record for this book
is available from the British Library

Library of Congress Cataloguing in Publication Data
Sustainable farm forestry in the tropics: social and economic analysis and policy / edited by S.R. Harrison, J.L. Herbohn.
p. cm. — (New horizons in environmental economics)
Includes bibliographical references.
1. Sustainable forestry—Tropics. 2. Tree farms—Tropics. 3. Sustainable forestry—Economic aspects—Tropics. 4. Tree farms—Economic aspects—Tropics. I. Harrison, S. R. (Stephen Robert), 1941– II. Herbohn, J.L., 1963– III. Series.

SD247.S87 2001
634.9'9'0913—dc21 2001023763

ISBN 1 84064 720 5

Printed and bound in Great Britain by MPG Books Ltd, Bodmin, Cornwall

Contents

Figures and boxes

Tables

Contributors

Ms M.L. Berry
Associate Lecturer in Accounting
Department of Accounting and Finance
James Cook University
Townsville QLD 4811
Australia

Dr S. Boydell
Reader in Land Management
Head, Land Management & Development
SSED, University of the South Pacific
PO Box 1168, Suva, Fiji.

Dr R.N. Byron
Commissioner, Productivity Commission
28th Floor, Telstra Tower
35 Collins St.
Melbourne VIC 3000
Australia

Mr M.A. Cox
Lecturer in International Business
School of Marketing and International Business
Faculty of Business
QUT Gardens Point Campus
GPO Box 2434
Brisbane QLD 4001, Australia

Mr C. Creighton
Executive Director
Australian Land and Water Resources Audit
GPO Box 2182
Canberra ACT 2601, Australia

Mr A.J. Cummine
National Policy Director
Australian Forest Growers
PO Box E18
Kingston ACT 2604
Australia

Mr N.E. Emtage
Research Officer
Cooperative Research Centre for Tropical Rainforest Ecology and Management
Cairns QLD 4350, Australia

Dr A.S. Ghose
Research Officer
Department of Economics
The University of Queensland
St Lucia QLD 4072
Australia

Mr K. Gould
General Manager Operations
DPI-Forestry
P.O. Box 944
Brisbane QLD 4001
Australia

Dr S.R. Harrison
Associate Professor of Economics
Department of Economics
The University of Queensland
St. Lucia QLD 4072
Australia

Dr J.L. Herbohn
Senior Lecturer in Natural Systems
School of Natural and Rural
Systems Management
The University of Queensland
Gatton QLD 4343
Australia

Mrs K.F. Herbohn
Lecturer in Accounting
Department of Accounting
The University of Queensland
St Lucia QLD 4072
Australia

Dr R.J. Keenan
Senior Scientist
Bureau of Resource Sciences
P.O. Box E11
Kingston ACT 2604
Australia

Dr D. Lamb
Associate Professor of Botany
Department of Botany
The University of Queensland
St Lucia QLD 4072
Australia

Dr M.J. Quayle
Senior Lecturer in International
Business
School of Marketing and
International Business
Faculty of Business
QUT Gardens Point Campus
GPO Box 2434
Brisbane QLD 4001, Australia

Dr M.E. Qureshi
Research Officer
Cooperative Research Centre for
Sustainable Sugar Production
James Cook University
Townsville QLD 4811
Australia

Mr D.B. Smorfitt
Lecturer Accounting and Finance
School of Business
James Cook University
P O Box 6811
Cairns QLD 4870
Australia

Ms N.H. Stehn
Project Leader, Economics
Environmental Policy and
Economics Division
Queensland Environmental
Protection Agency
160 Ann Street
Brisbane QLD 4000
Australia

Mr P.V. Townsend
PhD Student/Scholar
School of Resource Management and Environmental Science
Department of Forestry
The Australian National University,
Canberra ACT 0200
Australia

Mr T.J. Venn
Research Officer
Department of Economics
The University of Queensland
St Lucia QLD 4072
Australia

Dr S.M. Vize
Manager
North Queensland Afforestation Association Inc.
PO Box 2420
Cairns QLD 4870
Australia

Preface

There has been an upsurge of interest in small-scale forestry in tropical countries over the last 10 to 15 years, both for timber production and to achieve social and environmental objectives. This upsurge of interest has been associated with increasing community concern about environmental issues, in particular the negative effects of deforestation. Some of the world's finest timber species are endemic to tropical forests, with these forests being a major long-term economic and environmental resource if managed appropriately. However, many of the locations where tropical forests have existed have been in low-income and developing countries, which has exacerbated logging pressure to support livelihoods and generate export revenue. Even in more affluent tropical countries, there has historically been limited conservation effort. The rate of logging in most tropical countries has clearly been excessive, in retrospect, and governments have sometimes had limited ability to control deforestation. However, as the seriousness of deforestation has been recognized, logging concessions have been terminated, and residual forest has been accorded protected status, such as National Parks, Wildlife Parks and World Heritage Areas.

Strong imperatives exist for reforestation in tropical areas. Amongst these are the shortage of lumber and fuelwood, high timber import expenditure, the need to control land degradation in the uplands and lowland flooding and damage to hydropower systems, and excessive logging pressure on coconut plantations (for cocolumber). Forests in tropical countries, particularly in the wet tropics, are notable for their remarkable biodiversity, which once lost is not replaceable.

Against this background, there has been strong interest in establishing plantations and woodlots, often of non-traditional and non-coniferous species, for both timber and non-wood products as well as environmental services. This is in clear contrast with temperate forestry, with its continuing high reliance on a narrow range of conifer species, primarily for timber production. Forestry activity has included government, industrial, farm, social and community plantings. With increasing recognition of indigenous landrights, and in conjunction with other factors that have restricted access to large areas of land, government policy has tended to move away from large plantations towards supporting community and farm forestry.

The increasing importance of non-industrial forestry in the tropics raises research questions concerning aspects such as seedling technology, site-species matching, production systems, estimation of stand growth rates, marketing systems and predicting financial performance. Often, the system of property rights in place does not encourage small-scale forestry, and a host of other constraints can be identified, hence the need to examine measures to overcome impediments to more widespread planting.

This book, and a related volume (*Sustainable Small-scale Forestry: Socio-economic Analysis and Policy*, also published by Edward Elgar) examine some of these research issues. They arise as a result of the socio-economic program of the Cooperative Research Centre for Tropical Rainforest Ecology and Management (Rainforest CRC), a partnership of research organisations and research user organisations, with its headquarters based in Cairns in north Queensland, Australia. The socio-economic program forms an important component of this wide-ranging research thrust into improving understanding and management of tropical rainforests and reforestation with rainforest species. Some of the chapters arise from research carried out in north Queensland. As this work progressed, the editors came in contact with a wide range of people undertaking socio-economic research into small-scale forestry in tropical areas in other countries, and noted a high degree of similarity in issues of concern. As well, the editors have been successful in attracting support to carry out forestry research in other areas in Australia, the Philippines and India.

I commend Steve Harrison and John Herbohn, and the other contributors, for bringing together this interesting and insightful collection of research papers on small-scale forestry in the tropics.

Professor Nigel Stork
Director, Rainforest CRC
James Cook University, Cairns

SETTING THE SCENE

1. Evolution of Small-Scale Forestry in the Tropics

S. R. Harrison and J. L. Herbohn

The impetus for this book was the involvement of the authors in the Cooperative Research Centre for Tropical Rainforest Ecology and Management (Rainforest CRC), in North Queensland, Australia, and in forestry projects supported by the Australian Centre for International Agricultural Research (ACIAR), the Rural Industries Research and Development Corporation (RIRDC) and other agencies. The Rainforest CRC has provided a unique opportunity to examine reforestation within the context of a tropical high-rainfall, high-tourism area with a strong conservation lobby in a developed country. This CRC, and its socio-economics programme, commenced in 1993, about the time when 10 local governments obtained state and national government support for the Community Rainforest Reforestation Program (CRRP), in part as a means of replacing the timber resource lost by World Heritage Listing of the Wet Tropics of Queensland.

When we became involved in overseas projects, and with organisations such as the International Union of Forest Research Organisations (IUFRO) and the Centre for International Forestry Research (CIFOR), it became apparent that many of the issues facing small-scale forestry as an industry are common across various countries. Also, it soon became apparent that people involved in small-scale forestry have a different mindset to traditional foresters concerned with government and industrial plantations. Further, the dearth of documented research and general understanding of the characteristics and requirements of timber species grown in tropical small-scale forestry also became obvious. At the same time, we became aware that a great deal of research and experimentation is taking place with respect to small-scale forestry in the tropics and sub-tropics, by both researchers and landholders.

Within our own research programme, various socio-economic research projects and student dissertations have been conducted, generating a substantial amount of information, and it seemed timely to compile this information together. At the same time, we have sought to draw information from other tropical countries, and to arrange this on a thematic (rather than country) basis. While technical and economic concepts are not avoided, the material is written in a style designed to be readable by both professional and non-professional people with an interest in small-scale forestry. The book is organised into four sections, dealing with background setting, timber and marketing, non-wood forest benefits, and policy perspectives. This chapter introduces the themes of the book and outlines the section and chapter contents.

1.1 SMALL-SCALE FORESTRY IN THE TROPICS

The nature of forestry differs in the tropics compared with temperate regions. In the former – such as North America, Europe, the former Russian states and New Zealand – there tends to be a reliance on conifer species, often with long rotations. Where tropical forests are found – such as in equatorial Africa, Central and South America, Southeast Asia, northern Australia, Indonesia, Papua New Guinea and the Philippines – short rotation hardwoods are more common. The vast heterogeneity of locations and situations leads to huge variety in forestry systems and makes generalising about plantation forestry difficult. Many of these countries are so-called developing countries, with high population pressures and limited means to conserve their native forests.

While tropical regions were initially endowed with large areas of high quality timbers, recent rates of deforestation have been high. According to Sarre (1999, p. 1) of the International Tropical Timber Organisation (ITTO), 'the last century has been a torrid one for forests. They have been slashed, cut down, burnt, eroded and otherwise devastated like never before'. Often the economic rents from liquidation of forest capital appear to have been sequestered by a few large companies rather than well distributed across the community. Some former timber exporting countries, such as the Philippines, have become net timber importers. There has been a strong motive to establish plantation forests to provide a timber resource for residential and industrial purposes and to alleviate balance-of-payments problems. Accompanying deforestation, flooding of lowland areas and siltation of rivers has been exacerbated, providing another motive for reforestation. In some cases, plantation forestry is also seen as a means of reducing illegal logging and meeting fuelwood needs of the rural poor. In both developed and developing countries, plantation forestry can reduce the logging pressure on native forests, and assist in conservation of native biota.

What has become apparent in recent years is the difficulty in expanding government and industrial plantations. Land access has been a limiting factor, due to cost of acquiring land, hostility of local people to loss of access to what they have perceived as common property resources, land claims of indigenous people which have been difficult and time-consuming to resolve, and objections of environmentalists to clearing native vegetation for plantations. Sometimes tree survival rates in large plantations have been low, due to factors such as poor site-species matching, inadequate silviculture and wildfires.

Against this background, there has been steady progress in small-scale forestry, in which individuals and communities have chosen to devote land, capital and time to growing trees. In general, these activities can be classed as *farm* and *community* forestry. The latter may be distinguished from the earlier and generally discredited 'social forestry' by greater planning and management involvement by local communities (a 'bottom-up' planning style, as opposed to the 'top-down' planning style) and, normally, use of common property land resources. The high hopes of the 1980s for social forestry as a means of preventing illegal logging and providing fuelwood for the poor were in general not met, the beneficiaries being larger landholders and industrial foresters.

Small-scale plantings have ranged from a few trees to hundreds of hectares. In developing countries with high population density, plantings on field boundaries, roadsides and along canals is common. Sometimes mixtures of species are planted. Inclusion of fruit trees is popular, as a food source with some residual wood value. The standard of silviculture is often lower than for plantations managed by professional foresters. Both native and exotic species are grown in plantations in the tropics. Some species – such as mahogany, teak and gmelina – are widely grown. As well, Australian eucalypts and acacias play an important role in plantations in many countries, providing hardwood lumber, poles and woodchips.

While government and industrial plantations are normally designed for timber production as the dominant use, small-scale forestry may involve a wide range of objectives. In developing countries, production of fuelwood and non-wood forest products is often important. Protection of land and watersheds, landscape aesthetics and wildlife habitat are often important in more developed countries. Crop or stock protection with windbreaks and alley cropping are common.

The marketing of timber from small-scale forestry presents issues of low market power of sellers, and low volume, uniformity of product and continuity of supply. Often, returns to growers are very low, and constitute a small percentage of the final timber value. This has created impetus to formation of seller groups, such as timber cooperatives. In developing countries, much of the timber, fuelwood and non-wood forest products are used directly by producers, though well established markets often exist for these products.

1.2 THE ROLE OF GOVERNMENT IN SMALL-SCALE FORESTRY

Governments have played an important role in the development of small-scale forestry. In general, the private profitability of small-scale forestry tends to be low or uncertain. Forestry, however, provides positive externality benefits not captured by the grower, particularly in the form of environmental services such as land and water protection, wildlife habitat, landscape appearance and carbon sequestration. Forestry is an enterprise with a long payback period and relatively high risk. Without financial or other support, often landholders will not regard forestry as a worthwhile activity, so the area planted will be less than the socially optimal area. Hence there is a case for support from the taxpayer. In developing countries, forestry is often well regarded and targeted for support by international aid and loan agencies and non-government organisations (NGOs) as a means of facilitating economic and environmental sustainable regional development.

The role of government in small-scale forestry is not always clear-cut. Government often provides assistance to tree growers, such as subsidies, grants, provision of seedlings and extension services. At the same time, forestry departments also have an important regulatory role in protecting valuable ecosystems and habitats. For example there may be regulations preventing the harvesting of particular species, or timber on steep land or near watercourses. Further, unpredictable changes may be made in support levels, harvest rights, taxation regimes and so on, which reduce private profitability of plantations (a case of sovereign risk). Some government actions create uncertainty in property rights of landholders, and impede rather than promote plantation forestry. Hence the simplistic 'market failure government fix' model cannot always be relied upon, and there needs to be an awareness of the risk of 'regulatory failure' with respect to small-scale forestry.

1.3 THE NEED FOR RESEARCH INTO SMALL-SCALE FORESTRY

There is clearly a great scope for research to help fill the many knowledge gaps about small-scale forestry in the tropics, in terms of biology and the socio-economic aspects (the latter being the primary focus here). Identifying opportunities for forestry enterprises requires an understanding of the species or mixtures suitable for particular sites and their silviculture. Little is known about the growth performance of most of the species grown, which becomes a severe limitation for financial modelling. An understanding of the impediments to tree planting is critical if governments are to provide assistance. Some knowledge of landholder attitudes to and motives for tree planting, and of the forms of incentives to which they will respond, is also critical. Estimates of the returns from timber production and of the values of non-timber products and environmental services of plantation forestry are critical for policy support. Consideration of the timber production pipeline provides a focus for the analysis of costs of log production, harvesting, processing and marketing, and of rent sharing between the various stakeholder groups. Research into the attitudes and practices of timber users can provide guidance of species selection and timber quality requirements. Trade issues loom large when exports of logs or timber products are envisaged.

1.4 LAYOUT OF THE BOOK

This first section sets the scene for the book. Chapter 2 examines the history of forest management with reference to a case study in the Wet Tropics of Queensland, Australia while Chapter 3 reviews farm and other small-scale forestry systems in the tropics.

Section 2 is concerned with farm forestry from the perspective of timber industries, including forest management, timber processing and marketing. In general, there is a dearth of information about the potential profitability of small-scale forestry, and particularly profitability of mixed species plantings. Chapter 4 illustrates the problems of modelling returns from species mixtures, presenting a financial model for a mixture of two tropical species with alternative row planting and staged harvesting. Chapter 5 illustrates how a financial model may be used to estimate the investment risk in farm forestry due to yield and price uncertainty. In Chapter 6, a method is presented for estimating stand yield and financial returns for farm forestry in developing countries where little yield information is available.

Chapter 7 reports research into the role of portable sawmills in tropical small-scale forestry, and the economics of portable sawmills relative to fixed-site mills. Some issues concerning choice of timber inputs by furniture manufacturers and implications for species selection and timber marketing are introduced in Chapter 8. Chapter 9 examines issues in developing export markets for timber, including timber produced by small-scale forestry in tropical North Queensland.

Section 3 examines non-wood benefits of forestry, i.e. products and services apart from lumber production which are increasingly being recognized as important. Chapter 10 examines the nature and importance of these benefits, in developing and developed countries. Chapter 11 presents findings of a research project into the impact of reforestation with tropical rainforest species on the value of agricultural land. In Chapter 12, the findings of a further research project, on the application of multicriteria analysis to evaluate riparian revegetation options in the wet tropics, are presented. Chapter 13 examines the need and potential methods for inclusion of non-wood values in statutory accounting reports by agencies responsible for the management of forest resources, with particular reference to small-scale forestry in the tropics.

Section 4 takes a policy perspective and is concerned in particular with measures that may be taken to promote small-scale forestry in tropical areas. Taxation treatment can be an important factor in the private profitability or otherwise of forest plantations, and taxation provisions vary greatly between countries; in Chapter 14 the taxation arrangements for farm forestry in Australia are examined with reference to tropical small-scale forestry. Chapter 15 reviews the findings of a number of survey-based investigations into landholder attitudes to farm forestry, reasons for tree planting and government incentive programmes to promote tree planting. Chapter 16 examines factors that influence the uptake of farm forestry by smallholders in developing countries in the tropics. In Chapter 17, community forestry programmes are examined, with reference to joint forest management (JFM) in India and community based forest management (CBFM) in the Philippines. Institutional arrangements to encourage farm forestry are critically examined with respect to a 'community' forestry programme in northern Australia in Chapter 18. Chapter 19 examines arrangements governing stumpage prices in Australia, which are critical to the profitability of farm forestry. Future opportunities, impediments and challenges of small-scale forestry in the tropics are reviewed in Chapter 20.

REFERENCES

Sarre, A. (1999), 'New era, new thinking', *Tropical Forest Update*, **9**(4), p. 1.

2. Historical Background to Plantation Development in the Tropics: A North Queensland Case Study

D. Lamb, R.J. Keenan and K. Gould

This chapter provides a case study of the history of forest use and reforestation in the tropics, and the changing community attitudes to forestry over time, in a northern Australian setting. For more than 100 years North Queensland rainforests were viewed as a source of capital or as an impediment to agriculture. By the 1950s, much of the forest on easily accessible land with gentle topography was cleared. Some of this farming land was subsequently abandoned and it is now recognised that much of this clearing, especially on the uplands, was inappropriate and that the land should have remained under forest. New forms of silviculture were developed for the forests remaining on steeper lands and by the end of the 1980s satisfactory regeneration was occurring after logging. Although plantation trials were undertaken early in the 1900s, the apparently continuous supply of timber from natural forests was a disincentive to establishing larger plantation areas. The cessation of logging and a rise in interest amongst landowners about 'conservation' has led to an opportunity to reforest some of the cleared lands. A major research effort will be needed to develop appropriate forms of plantation silviculture for the region.

2.1 INTRODUCTION

This chapter reviews the history of forest use in North Queensland, to provide a background context for the present interest in reforestation. In a little over 100 years forest land use in North Queensland has passed through three contrasting stages, i.e. pre-1890, the period from 1890 to 1988 and 1988 onwards. Prior to 1890, there was unconstrained timber exploitation by timber fellers seeking the highly-valued Red Cedar (*Toona ciliata*). The second stage involved a long period of conflict between agriculturalists wishing to clear forests and those wishing to retain forests for either timber production or nature reserves. The over-riding land use policy that evolved in this time was for closer agricultural settlement and much of the forest on flatter land was cleared. There was an unwillingness to accept forestry as an essential rather than residual land use (Frawley, 1985). The final stage has coincided with the recognition that much of the earlier clearing was over-zealous and attempts have been made to preserve the remaining forests and to reforest some previously cleared land.

The human cost of this early forest clearing was high; the work was difficult and many of the farms created from the former rainforest areas eventually failed. It is now apparent that the financial cost of reforestation of these cleared lands will also be high. The irony is that some of those involved in the early forest clearings have lived to see others working to reverse the results of their labour.

This chapter briefly outlines the history of forestry and forest land use in northern Queensland to provide a background to this reforestation effort. The material here draws heavily on the reports of Birtles (1988), Carron (1985), Frawley (1988), and Taylor (1994), and more detailed accounts of general forest-use issues are available in these publications.

2.2 THE EARLY YEARS OF EUROPEAN SETTLEMENT

When the colony of Queensland was separated from New South Wales in 1859 there was limited European settlement in the north. In 1862 the botanist Walter Hill visited North Queensland and commented on the luxuriance of the 'scrub' (i.e. rainforest) around the Tully River (Birtles, 1988). Timber cutters in NSW and southern Queensland were already familiar with species such as Red Cedar (*Toona ciliata*) and Hoop Pine (*Araucaria cunninghamii*) and Hill's report attracted cedar cutters to the north. Others were soon attracted by gold and tin mining. From the mid 1870s cedar cutting had begun in many of the northern river valleys. Fears arose that uncontrolled forest clearing might affect rainfall and in 1870 the Acclimatisation Society of Queensland wrote to the Colonial Secretary expressing concern (Carron, 1985, Taylor, 1994). Interestingly, this same issue was raised in other British colonies at the time (Dawkins, 1997). Further letters followed without any government response. The Society subsequently held a public meeting in Brisbane to air its concerns but one of those speaking (L.A. Bernays) subsequently wrote:

> The idea of a scarcity of timber in Queensland is often met with a smile of pity or even a laugh of derision. One is asked to get upon any eminence, in any district, and to put oneself the question – whether the sea of timber that meets the eye can ever run short? (Taylor, 1994).

Eventually the government did appoint a Select Committee of Parliament to consider the issue further but early timber cutting remained essentially unregulated and by 1880 most of the coastal valleys had been cleared of Red Cedar. Inevitably the cutters were forced to turn to other species and by the late 1880s the value of species such as Kauri Pine (*Agathis robusta*), Black Bean (*Castanospermum australe*), Beech (*Gmelina fasciculiflora*), Northern Silky Oak (*Cardwellia sublimis*), Queensland Maple (*Flindersia brayleyana*) and Maple Silkwood (*Flindersia pimenteliana*) were being recognised.

In 1880, W.A. Tully (Under-Secretary for Lands) noted in a report to the Legislative Assembly that logging was proceeding faster than the capacity of the forest to regenerate (Birtles, 1988). He attempted unsuccessfully to develop a 'forest conservancy' to protect some of the remaining cedar stands. The particular difficulty for those wishing to limit clearing of rainforests in North Queensland was that these rainforests were seen to be indicative of fertile soils. By the early 1880s sugar cane was becoming established on flat lands on the coast and such fertile lands were in high demand. While there was a similar enthusiasm for clearing forests in the cooler climates on the uplands around Atherton, there was less certainty over the most appropriate forms of agriculture to practice. Eventually dairying became established and large areas of the tablelands were cleared. Data provided by Easterby (1932) suggests it took about 10 man weeks of work per hectare to fell, burn and clear rainforest on the coast at the turn of the century. Methods improved and Smith (1991) reports that a skilled axeman could later fell about 0.4 ha a day using the 'drive' technique in which a single tree would be deliberately felled into a number of partially cut trees causing a large area to be felled at once. High rates of land clearing meant large timber volumes were available and the prices of even the best timbers were low, with many of even the best trees simply burned. In commenting on these practices the Under-Secretary of Department of Lands noted 'maintenance of these rich volcanic coastal scrubs as permanent reservations for forestry purposes cannot be regarded as a subject for serious consideration' but he also commented 'how far the excellent indignous pines of the State can be reproduced on soils inferior in quality to that in which they are now found to be naturally produced is a question yet to be determined' (Carron, 1985).

The first serious attempts to halt wasteful clearing and to regulate forest use occurred in 1906 with the passage of the *State Forests and National Parks Act* and establishment of a Forestry Branch. The Branch had three staff to administer forestry operations over the whole state and although it attempted to reduce the profligacy of the various land clearing operations it was out of step with the political realities of the time. Only when E.H.F.

Swain was appointed as Director of Forests in 1918 did the Branch acquire a spokesman with the self-confidence and force of personality able to demand a more rational pattern of land use. Following the end of the First World War there had been pressure for large-scale clearing for agriculture between Cooktown and Tully without any prior assessment of land capability or markets. These proposals were vigorously opposed by Swain who pointed out the high value of timbers in these forests and the poor use made of previously cleared forest lands. In Swain's view:

> The Lands Administration of the day disclosed itself as merely a wholesale land opening agency, having no perception of the development importance of natural resources, no appreciation of conservation principles, no idea of land economics or of land utilisation, no policy at all except to parcel out land into individual blocks regardless of the consequences . . . (Frawley, 1985).

Swain urged that more vacant crown land in the north be incorporated in State Forest or National Park reserve. In Swain's view forestry was also land settlement and forests could be managed in perpetuity to provided local employment and economic benefits. He was opposed by those in the Lands Administration who favoured more agricultural clearing and by many politicians. The issue came to a head in 1929 with the 'Three Million Pound Scheme' which was a proposal to clear 145,000ha of forest for agricultural development between Millaa Millaa and Innisfail. Swain's vigorous and public opposition to the scheme led in 1931 to a Royal Commission to investigate on land clearing, land use and forestry administration. The Commission recommended further land clearing for agricultural purposes, and urged the Government to fund research on commercial agricultural crops for the region. It also recommended a reduction in public expenditure on forestry administration while, at the same time, recommending an increase in the area of forest reserves. Swain was effectively dismissed when the Forestry Board he chaired was replaced by a Sub-Department of Forestry within the Lands Administration Board. In the Commission's view, 'Queensland needs no forest science for present requirements . . . the productive wealth of the country at present suffers from the fact that there are too many, rather than too few trees' (Payne *et al.*, 1931). The removal of Swain seriously weakened forestry interests within the Public Service in Queensland for many decades.

The clearing of rainforested lands for agriculture continued on the tablelands until the 1950s by which time it was apparent that the remaining forested land was mostly unsuitable for agriculture. Small areas of land clearing for agriculture have continued on coastal areas to this day.

In summary, the early years of European settlement in North Queensland were marked by a long period of unregulated forest harvesting followed by an even longer period when government policy favoured rapid clearing for agriculture. The general public were mostly in favour of this program of deforestation since forestry and nature conservation were seen as residual rather than essential landuses. This was despite the high value of many of the rainforest timbers and the unique diversity of the biota in these northern forests.

2.3 EARLY AGRICULTURAL EXPERIENCES IN NORTH QUEENSLAND

Most of the early agricultural development in north Queensland took place on the flatter coastal areas and this is where most rainforest clearing took place. Winter *et al.* (1987) estimated that about 57% of rainforest land below 300m elevation was cleared, in comparison with only 5% of the foothills and 14% of the rainforest lands above 300m elevation.

While a variety of crops were tested on the coastal lowlands – including cotton, coffee, maize, tobacco and rice – sugar became established as the pre-eminent agricultural crop (Bolton, 1963). There was a ready market and prices were mostly high. Sugar growing commenced in NSW and southern Queensland but soon spread north and by the mid 1880s there were 166 mills in operation in Queensland with North Queensland producing two thirds of the Queensland crop (Bell, 1956; Bolton, 1963). Once the success of sugar was established many of the subsequent plantings were carried out by companies which established estates around company-owned mills. Over time many of these company-owned plantations were broken up into smaller family farms (Easterby, 1932; Bell; 1956).

Sugar grew well in the climatic conditions of the coast, particularly on the more fertile 'scrub' soils. It soon became clear, however, that production levels could not be sustained. Harvests declined from 100–150 t/ha to less than 50 t/ha due to a loss of nutrients in harvesting and a lack of fertilising (Maxwell, 1900). A number of pests and diseases also became common in many areas. Because of the economic and social significance of the industry, the government responded by establishing the Bureau of Sugar Experimental Stations in 1900 (Easterby, 1932). The Bureau subsequently carried out a detailed and extensive research program to boost yields and place the industry on a more stable and productive footing (Bureau of Sugar Experimental Stations, 1950). This involved development of fertiliser schedules, controls for pests and diseases and improvements in the genetic base of planting stock. The sugar industry is now one of the dominant agricultural industries of tropical Australia.

Agricultural development on the tablelands was slower to develop despite a high degree of enthusiasm. Although 50,000ha had been thrown open for settlement on the tablelands in 1882 there were difficulties in clearing land and establishing farms and by 1890 only 667ha had been cleared and cultivated. Most was used for maize. Paspalum grass (*Paspalum dilatatum*) was introduced after 1900 and by 1913 the area under pasture exceeded that under maize (Frawley, 1987). A butter factory was established and dairying became the dominant landuse (Bolton, 1963).

As on the coast there was a widely held view that all 'scrub' soils were fertile but it was evident that these soils varied across the tablelands due to differences in underlying geology and rainfall. The basaltic soils were more fertile than the granitic soils and fertility was highest in the lower rainfall areas in the north west of the tablelands where leaching was less than in the wetter, southern areas. This range of conditions led to cropping being favoured in the northern areas and dairying covering most of the wetter, southern areas. The variations also meant it was difficult to prescribe a single method of establishing pastures and a range of prescriptions was developed for different areas. For example, Paspalum was common around Malanda but Kikuyu (*Pennisetum clandestinum*) was used in the wetter areas around Millaa Millaa. In the nearby Maalan area government agronomists recommended settlers use Molasses Grass (*Melinus minutifloria*) but dry winters and frosts caused the pastures to die off and much trial-and-error planting of various grass species had to be carried out by settlers themselves to find an alternative (Frawley, 1987).

As on the coast, the fertility declined rapidly after farming commenced (Straughan, 1947; Sloan, 1956). The decline was most severe in the lands used for dairying and many farmers found they were unable to sustain high quality pasture species or support as many stock as they originally had. Sloan (1956) reported stocking rates dropping from one animal per acre (0.4 ha) in the early days to one animal to four or five acres (*ca.* 2ha) by the mid 1950s. Trials to overcome these declines using fertilisers, and later tropical legumes, began in the late 1940s. Again trial-and-error plantings over several decades eventually led to the conclusion that Tinaroo Glycine (*Glycine wightii* cv. Tinaroo) was suitable in the drier areas around Atherton and Kairi but failed in wetter areas to the south and east of Malanda where *Desmodium* spp. were more suitable. In each case a separate fertiliser regime was needed.

Forest clearing on the tablelands largely ceased in the early 1950s. In 1970 a review of land use across the region found that many dairy farms were barely profitable and that about 59,000ha of land was either degraded or had been abandoned and was occupied by forest regrowth (Gilmour and Reilly, 1970). Even the most recently cleared land at Maalan, south of

Millaa Millaa that had been settled only 20 years earlier was mostly classed as degraded despite the many years of accumulated agricultural experience available at the time it was settled (Frawley, 1987). It was evident that government and the public enthusiasm for 'development' had pushed forest clearing into areas quite unsuitable for agriculture with high social and ecological costs.

Agricultural development on former rainforest lands therefore differed on the coastal areas and the cooler uplands. On the coast, sugar quickly became the dominant crop because it grew well and because there was a mostly large and profitable market. Large companies were an early component of the industry and there was little government research support until about 40 years after sugar was first planted. On the tablelands there was also vigorous agricultural activity but success was less assured. Much trial and error was necessary before pasture mixes and management methods were developed and agricultural activities were sometimes extended onto areas unsuitable for pasture development or cropping.

2.4 FORESTRY AND PLANTATION DEVELOPMENT

The new Forestry administration faced severe problems when it was first established in 1900. Most of the forest estate was remote and difficult to access, many of the forest trees were unknown to science and the silvicultural requirements of the 10 commercially attractive rainforest species then being used were poorly understood. The first tasks of the new administration were to try to bring some order and management to the essentially unregulated industry and to secure a permanent forest estate. There were particular concerns at the turn of the century that supplies of the highly valued Red Cedar were shrinking so that trials with this species were carried out at Wongabel near Atherton in 1903 using wildlings (i.e. transplanting naturally occurring seedlings). These trials were abandoned in 1906. Red Cedar, like many members of the Meliaceae, is bedevilled by a shoot tip borer (*Hypsipyla*) and the early results must have been discouraging. Besides, large volumes of timber were being felled and burnt in the land clearing frenzy then underway on the tablelands. It was little wonder that work on reforestation seemed so pointless.

Further plantation and underplanting trials with a larger number of species were carried out at Wongabel in 1912. The results were generally disappointing because of damage from animal browsing and insect attack but the work identified Hoop Pine (*Araucaria cunninghamii*) and Kauri Pine (*Agathis robusta*) as promising new plantation species although the former species was not native to the Atherton tablelands. But again the trials were abandoned when the new Director of Forestry, N.W. Jolly decided that expenditure on plantations would not be warranted and that attention should be directed at assisting regeneration in logged forests.

Forestry in Queensland was boosted by the appointment of E.H.F. Swain as the new Director of Forests in 1918. He initiated a number of developments including research into the utilisation of a wider range of Queensland species. This resulted in a landmark publication on the properties of native timbers including rainforest species. He also devised new methods of charging timber royalties that allowed forests more distant from sawmills to be harvested as profitably as those closer to markets. Swain saw North Queensland cabinet timbers as being a particularly valuable resource and work was carried out on enrichment planting using Queensland Maple (*Flindersia brayleyana*) and Maple Silkwood (*Flindersia pimenteliana*). He also promoted plantation development using mainly Hoop Pine and Kauri Pine to counter what he saw as a future shortfall in the supply of native softwoods.

Plantation development continued on the tablelands after Swain's departure and several hundred hectares of highly productive plantations were established before the program was suspended with the onset of World War II. The sawmilling and plywood industry also developed significantly during this period. Until 1936 there were no restrictions on the establishment of sawmills or on the quantity of timber a mill could cut. Given the large amount of agricultural clearing then taking place in north Queensland it would have been pointless to do so.

Logging in native forests expanded rapidly during World War II. After the war there was a substantial increase in the size of the harvest in reserved forest and also in the number of species utilised (from about 36 species in 1940 to a final total of more than 124 species). This was accompanied by the recognition that not all rainforest timbers were of equal value and that there was a need to develop various timber quality classes with matching royalty charges; eventually four timber quality classes were recognised.

Planting also recommenced on the tablelands after the war. Throughout Australia there was an increase in softwood plantation development and large areas of exotic pines were established in most states with the support of federal funds. In Queensland, large plantations of *Pinus elliottii* and *Pinus caribaea* were planted in the south near Brisbane on lands regarded as being too infertile for agriculture. In the north, plantations were established on farmlands that had been resumed by the government at the time of the construction of the Tinaroo Dam. In places Hoop Pine was used but the more tolerant *Pinus caribaea* was also planted on some of the less fertile soils. Plantations were also established on some of the less fertile soils on the coast. In total around 13,000ha were established.

Once again plantation trials using rainforest species were carried out, using many of the species found in the higher timber quality classes noted above, but the growth of most of these species was slow (< 5 m^3/ha/yr) in comparison with the productivity achievable with exotic pines (around 15 m^3/ha/yr) and Hoop Pine (*Araucaria cunninghamii*) remained the only rainforest species widely used in commercial plantations. By the 1990s about 50,000ha of this species had been established in North and South Queensland. A noticeable feature of all of this reforestation activity was that it was mostly initiated by the government. Unlike the sugar industry there were no privately managed forest estates or plantations except for an exotic pine plantation established by a paper company in southern Queensland. The long investment periods required in plantation forestry and the deliberate government policy of maintaining artificially low timber prices were major disincentives for investors. Besides, there was still a widely held view that sufficient high-quality timbers were available from the natural rainforests and that regeneration in these forests after logging was sufficient to provide for future supplies.

Throughout the post-war period there was an increase in community interest in forests and conservation throughout Australia, particularly in the urban centres. A number of controversies emerged over land-use planning and forest management. Many of these were local in scope but several eventually involved conflicts between state governments and the federal government. These included mining on Fraser Island (1976) and the damming of the Franklin River in Tasmania (1983). In the late 1980s concerns began to be raised by conservationists over logging in the rainforests of North Queensland. This was prompted by the remarkably high diversity of plant and animal taxa contained in these forests despite them covering only a small area of the continent. The conflict eventually became a disagreement between the federal government which supported the cessation of logging and the Queensland state government which argued it should continue. The issue was resolved in 1988 when the area was listed on UNESCO's World Heritage register and logging ceased (Adam, 1992; Taylor, 1994; Vanclay, 1996).

The decision to cease logging had several consequences. These included the immediate loss of jobs in the timber industry and the almost complete reduction in the supply of high-quality cabinet timbers from North Queensland. While both state and federal governments considered compensation schemes to help displaced workers, it also seemed opportune to revisit the earlier idea of establishing plantations of some of the high-value native timber species that had been first considered more than 90 years earlier. If the real value of these timbers could be obtained from the market perhaps the slower growth and commercial unattractiveness of rainforest plantations could be overcome? Further, the softwood plantation programme had led to improved techniques in plantations establishment. Perhaps better growth rates could be achieved with rainforest species using modern plantation methods?

The scientific database on which to base a new plantation programme using rainforest species was patchy. The timber properties of the main species were now known and their relative value in the market was established but the silvicultural techniques needed to establish them in plantations was less well known. Similarly there was much uncertainty about the growth rates that might be possible in different areas. An important review of previous Forest Service trials was reported by Cameron and Jermyn (1991). This suggested a number of species showed promise and recommended further study of these. Economic projections by Shea (1992) and Russell *et al.* (1993) suggested plantations of high-value rainforest species could be profitable despite these uncertainties.

In 1992 the federal and Queensland governments with the local governments in North Queensland decided to fund the Community Rainforest Reforestation Program. The purpose of the programme was to foster tree planting on privately owned land throughout North Queensland in order to assist unemployed people in the region and to help recreate a plantation-based timber industry. But old attitudes change only slowly. While many landowners in North Queensland were keen to participate in tree planting for commercial or conservation purposes others were not and regarded trees as impediments to agricultural development in the same way that their forebears had over 100 years earlier. At a meeting of the Atherton Shire Council in October 1998 many Councillors were hostile to the idea of timber tree plantings on flatter lands in the shire. Tree plantations were seen as a source of pests (e.g. rats, foxes) and earlier farm plantings of the (low quality) *Pinus caribaea* were viewed as economic failures. The Council clearly saw forestry as a residual land use that should be restricted to the more hilly land unsuitable for agriculture. The traditional views concerning trees and forests that had dominated land-use planning in the north still held sway in at least some parts of the community.

2.5 CONCLUSIONS

North Queensland rainforests are a long way from the state capital of Brisbane, and remote from markets and public attention. For much of the time since European settlement, these forests have been seen as a hindrance to economic progress rather than as an asset to be safeguarded. These views have changed among some, but by no means all, members of the community. In the early days, negative perceptions were accompanied by uncertainties over how to manage remaining forests for sustained timber production and watershed protection. Over time, much has been learnt about the ecology of these species-rich forests and silvicultural systems have been developed to manage them. However, the focus of landowners on agriculture rather than forestry and the supply of timbers from natural forests meant that plantations of slow-growing rainforest trees were an unattractive form of investment. When governments invested in plantation forestry, the most politically and economically attractive sites were land close to markets in the south-east of the state.

Much is now known about the ecology of the rainforests of North Queensland but comparatively little is known about restoring these on cleared land or about establishing plantations of high-value native tree species for economic purposes. When agricultural systems were developed in North Queensland there was considerable trial and error involved in choosing pasture and crop species and management methods. Most of this was done by individual landowners because they could envisage a financial return. A similar period of trial and error will probably be necessary in the case of plantation forestry. The difference is that few landowners have the ability to invest over the time frames necessary to develop appropriate silvicultural systems and long-term government support will be essential.

REFERENCES

Adam, P. (1992), *Australian Rainforests*, Oxford: Clarendon Press.

Bell, A.F. (1956), *The Story of the Sugar Industry in Queensland*, Brisbane: University of Queensland Press.

Birtles, T.G. (1988), 'European interpretation of the Atherton-Evelyn "Vine Scrub" of tropical North Queensland, 1880-1920', in J. Dargavel, K. Dixon and N. Semple (eds), *Changing Tropical Forests: Historical Perspectives on Today's Challenges in Asia, Australasia and Oceania*, Canberra: Centre for Resource and Environmental Studies, Australian National University, pp. 197-218.

Bolton, G.C. (1963), *A Thousand Miles Away: a History of North Queensland*, Canberra: Australian National University Press.

Bureau of Sugar Experimental Stations (1950), *Fifty Years of Scientific Progress: a historical review of the half century since the foundation of the Bureau of Sugar Experimental Stations*, Brisbane: Government Printer.

Cameron, D.M. and Jermyn, D. (1991), *Review of Plantation Performance of High Value Rainforest Species,* Canberra: Australian Centre for International Agricultural Research.

Carron, L.T. (1985), *A History of Forestry in Australia*, Canberra: Australian National University Press.

Dawkins, C. (1997), *Silviculture in the Tropical Rainforest: an historical analysis of success and failure,* Oxford: Oxford Forestry Institute.

Easterby, H. (1932), *The Queensland Sugar Industry: a Historical Review*, Brisbane: Government Printer.

Gilmour, D.A and Reilly, J. (1970), 'Productivity survey of the Atherton Tableland and suggested land use changes', *Journal of Australian Institute of Agricultural Sciences*, **36**, pp.259-272.

Frawley, K. (1985), *Past rainforest management in north east Queensland: a background to the Daintree road issue*, CRES Working Paper 1985/27, Canberra: Centre for Resource and Environmental Studies, Australian National University.

Frawley, K. (1987), *The Maalan Group Settlement North Queensland 1954: A Historical Geography*, Monograph Series No.2, Canberra: Department of Geography and Oceanography, Australian Defence Force Academy.

Frawley, K. (1988), 'Exploitation to preservation: appraisals of the rainforest lands of north east Queensland, Australia', in J. Dargavel, K. Dixon and N. Semple (eds), *Changing Tropical Forests: Historical Perspectives on Today's Challenges in Asia, Australasia and Oceania*, Canberra: Centre for Resource and Environmental Studies, Australian National University, pp. 181-196.

Maxwell, W. (1900), *Report upon an investigation into the condition of the sugar industry in Queensland*, Brisbane: Government Printer.

Payne, W.L., Campbell W.W. and Atherton J.G. (1931), *Report of the Royal Commission on the Development of North Queensland (Land Settlement and Forestry)*, Brisbane: Government Printer.

Russell, J., Cameron, D.M., Whan, I.F., Beech D.F., Prestwidge, D.B. and Rance S.J. (1993), 'Rainforest trees as a new crop for Australia', *Forest Ecology and Management*, **60**, pp. 41-58.

Shea, J. (1992), *New Timber Industry Based on Valuable Cabinet Woods and Hardwoods*, Consultancy report for Councils of the Wet Tropics Region, Brisbane: Queensland Forest Service.

Sloan, W. (1956), 'Some aspects of land settlement in rainforest areas of north Queensland', *Journal of Australian Institute of Agricultural Science*, **22**, pp.129-136

Smith, L.W. (1991), *The Trees that Fell*, Ravenshoe: Privately Published, L. Smith.

Straughan, W.R. (1947), 'Problems of settlement on northern tablelands', *Queensland Agricultural Journal*, **64**, pp. 133-138.

Taylor, P. (1994), *Growing Up: Forestry in Queensland*, Brisbane: Allen and Unwin, p. 242.

Vanclay, J. (1996), 'Lessons from the Queensland rainforests: steps towards sustainability', *Journal of Sustainable Forestry*, **3**, pp. 1-25.

Winter, J.W., Bell, F.C., Pahl, L.I. and Atherton, R.G. (1987), Rainforest clearing in northeastern Australia, *Proceedings of Royal Society of Queensland*, **98**, pp.41-57.

3. Silvicultural Research and Development of New Plantation Systems Using Rainforest Tree Species

D. Lamb and R.J. Keenan[1]

Species choice, plantation design and management methods all influence the economic and social returns from plantation forests. Although Teak (*Tectona grandis*) has probably been grown in plantations in the tropics for over a 100 years it is only recently that serious efforts have been made in many tropical regions to undertake silvicultural research and develop appropriate plantation systems for other native rainforest hardwoods. This is in contrast with temperate regions where there is a long tradition of plantation silvicultural research.

Some of this temperate experience will be transferable but much will be inappropriate. The diversity of potential plantation tree species, the higher levels of both photosynthesis and respiration, the longer growing seasons and the faster rates of nutrient turnover all mean that plantations in the humid tropics are likely to require new types of plantations, new forms of silviculture and new methods of management. Social and economic circumstances may also mean that a different range of products and ecological services are required from these plantations.

Considerable time will be needed to develop appropriate silvicultural systems as well as new plantation designs and to develop appropriate ways of providing the mix of products and environmental services required for tropical regions. This chapter discusses the silvicultural research and the development of plantation systems that are suited to tropical areas. The research that has been undertaken in this area in North Queensland, along with research conducted elsewhere in the world, is used to illustrate some of the opportunities and problems.

---

[1] We would particularly like to thank Mark Annandale, Geoff Borschmann, Mila Bristow, Bob Congdon, David Doley, Nahn Duc Huynh, Ken Robson, Gary Sexton, David Yates, Oliver Woldring, previous Queensland Forest Service staff, and postgraduate students in the Cooperative Research Centre for Tropical Rainforest Ecology and Management for their contributions to various aspects of this work.

3.1 CHOOSING SPECIES FOR TROPICAL PLANTATIONS

Species choice is probably the most important issue in a plantation program. Commercially successful plantation tree species usually have several attributes, including rapid growth under plantation conditions, good form (i.e. straight stem with limited branching) and appropriate wood quality for the particular end uses (Palmer, 1992). Good plantation species should also be tolerant of a wide variety of soils and site conditions and be resistant to pests and diseases. In the particular case of plantations based on rainforest species in North Queensland, an appropriate species would also need a high commercial value to compensate for the low growth rates relative to exotic conifers and greater distances from major timber markets.

Despite the large diversity of trees in tropical forests and extensive experimentation and species testing throughout the world, the majority of tropical timber plantations involve monocultures of just three genera, *Pinus, Acacia* and *Eucalyptus*, with smaller areas of *Tectona grandis* and *Swietenia* spp. Even within these genera, most of which contain a large number of species, only a comparatively small number of species have been used extensively. A large amount of research over the last 30-40 years has produced nursery technologies and silvicultural prescriptions for raising, establishing and managing plantations of these species. Provenance testing and tree improvement has improved tree growth and form. Further developments include cloning of the best individuals and the creation of hybrids with improved properties from both parent species. For example, hybrids developed in sub-tropical Queensland in recent years include *P. caribaea* x *elliotii* and *E. grandis* x *urophylla* (Nikles and Lee, 1998).

In tropical areas, such as North Queensland, one option would be to follow the traditional pattern for tropical plantations and establish monocultures of pines, acacias, eucalypts or hoop pine (Table 3.1). The chances of economic success would be relatively high. Large plantations of these already exist in southern Queensland and in temperate Australia. The first three of these species are also widely planted in other tropical areas, particularly in S.E. Asia and South America. All have a relatively high degree of market acceptance and silvicultural techniques have been reasonably well established through long-term research programmes, though this is less the case with the tropical eucalypts (Francis and Shea, 1991; Lewty and Last, 1998). Such plantations might appeal to landowners primarily concerned with economic outcomes. However, they do not supply the new niche market for high-value cabinet timbers and are unlikely to be attractive to those interested in combining production and conservation benefits let alone those mostly interested in conservation plantings.

Plantations of rainforest species carry more risk because less is known about their silviculture and management. North Queensland rainforests contain over 100 commercially useful tree species. Trials have been undertaken with a number of these but many were initiated several decades ago when plantation establishment techniques were not as developed as they are now. Further, with the exception of *Araucaria cunninghamii*, most of these trials have been limited to assessing growth rates and not with nutrition, tree breeding or stand management issues such as thinning or pruning (Dale and Johnson, 1991). In developing a small-scale plantation program with rainforest species in North Queensland it is clearly impossible to match the research effort already undertaken on tropical eucalypts, pines or acacias. On the other hand, there is a good deal now known about the attributes and market value of most North Queensland timber species as a legacy of previous utilisation and there is some knowledge about growth rates and silvicultural requirements of many of these species in natural forest conditions (Vanclay, 1994). Less is known, however, about nursery techniques for raising large numbers of seedlings, or the management required for optimum performance. Some have been tested in earlier plantation trials but more recent methods of establishment, weed control and fertilisation might provide improvement on previously observed poor performance (Cameron and Jermyn, 1991).

Table 3.1 Alternative plantation options for North Queensland

Plantation type	Prodn. potential	Rotn. length	Existing silvicultural knowledge	Timber value	Conser-vation value	Ease of manage-ment
Caribbean pine	High	Short	High	Low-medium	Low	Easy
Eucalypts	High	Short	Moderate	Medium	Low-medium	Easy
Hoop pine	Moderate-high	Long	High	Medium	Medium	Easy-moderate
Rainforest spp monoculture		Long	Low	Medium-high	Medium	Moderate
Rainforest spp. mixtures	Low-moderate	Long	Low	Medium-high	High	Moderate-difficult

The Community Rainforest Reforestation Program (CRRP), established in 1992, sought to use only tree species from North Queensland and a large variety of these were established at a number of sites. As well as local rainforest species, several acacia and eucalypt species from the region were included. Many of these plantations are now more than six years old, and growth is being monitored in about 100 plots on farmland sites across high rainfall areas of North Queensland. Early growth in some plantations has been rapid, and annual average height growth of over 3m has been achieved on some sites. The best performing species in terms of early height growth are *Acacia mangium* and a number of eucalypt species (*E. pellita*, *E. cloeziana*, *E. grandis* and *E. camaldulensis*). Rainforest species that are growing well include *Elaeocarpus grandis*, *Flindersia brayleyana* and *Melia azerderach*. Growth is generally more rapid on lower elevation sites with high rainfall and reasonable soils. Height growth decreases rapidly with altitude and with decreasing moisture availability, although these responses vary considerably between species (Bristow, 1996). Further evaluation will be needed as these species age to understand their performance at different sites over time and if possible to reduce the variety of potential plantation species to a more manageable number.

3.4 PLANTATION DESIGN

Tropical plantation forestry can take a variety of forms (Evans, 1992). Most commercial timber plantations are established using a single species, commonly planted in regularly spaced rows. The site is generally cultivated in intensively managed commercial plantations, often using a winged ripper to aerate the soil and provide a greater effective rooting depth. Weeds are controlled using chemical herbicides before and after planting until canopy closure excludes most weeds. Initial planting densities range from 500 to 1500 stems per hectare (sph) with higher densities preferred in areas where rapid site capture is required. High densities are also an advantage where trees have been raised from seeds collected from natural forests rather than specialised seed orchards and where heavy early thinning of poorly-formed or slow growing individuals will be required. In essence, the trees are treated like any other agricultural crop. Such a plantation design can be used for a variety of species, including many rainforest species, and would be appropriate for landowners with a commercial focus.

Many landowners in North Queensland, however, are interested in plantation systems with a 'conservation' benefit as well as providing just timber production. For these landowners mixed species plantations are more appealing than plantation monocultures. Planting tropical rainforest species in mixed species plantations is an intuitively attractive idea. Apart from their obvious aesthetic and biodiversity benefits, mixtures might simulate at least part of the environment normally experienced by most rainforest tree species. Mixtures can consist of two or more species planted at the same time, or planting times can be staggered with a 'nurse crop' planted first to provide cover for the commercially favoured species, which is planted after the 'nurse' species is established. This method can be useful when young seedlings of a species are unable to tolerate full sunlight and need some early shelter to become established.

There are theoretical reasons for expecting productivity to be higher in mixtures than monocultures – some of which are set out in Table 3.2 – and mixtures have been tried by tropical foresters in a number of locations (Wormald, 1992). However, there are remarkably few replicated, long-term trials to judge the field performance of plantation mixtures and plantation monocultures. The primary reason for this is that mixtures are difficult to design and manage. Random assemblages of otherwise attractive plantation species are unlikely to be commercially successful. One species often grows faster and shades out the slower-growing species. Faster growing species may be ready for harvesting earlier but the removal of this species may damage the slower growing trees. There are sufficient examples of successful mixed-species plantations to demonstrate that they are feasible, but success usually requires rather more ecological and silvicultural knowledge about a range of species than is commonly available. This has led many foresters charged with developing productive forests to be sceptical about the practical value of mixtures (Wadsworth, 1998).

Despite this scepticism there is evidence from a number of well designed field trials demonstrating that particular types of mixture can have higher productivity than monocultures of the same species. The most common of these is where the mixture contains a nitrogen fixing species such as albizia or acacia species able to supply additional nitrogen at sites where the plantation would otherwise be nitrogen-deficient (Binkley, 1992; Binkley and Ryan, 1998). In other situations productivity may be enhanced by reduced insect damage in mixed species plantations relative to monocultures (Montagnini *et al.*, 1995). In North Queensland a long-term mixed species study found a lower incidence of infestation by the tip borer (*Hypsipila robusta*) when the valuable rainforest timber species red cedar (*Toona ciliata;* Meliaceae) was planted under the cover of *Grevillea robusta* compared with open planted trees (Keenan *et al.*, 1995). The reason for the reduced attack is not yet clear but the study clearly revealed that mixtures can be a way of raising this highly valuable timber species in plantations.

Table 3.2 Some potential advantages and disadvantages of using more than one tree species in a plantation

Potential advantages of mixture	Comment
More 'natural', simulates ecology of natural forest.	Many -perhaps most – species from diverse forests able to grow successfully in plantation monocultures.
Increased plant biodiversity.	Any increase is limited - most plantation mixtures only contain 2-4 species.
Increased productivity because of reduced between-tree competition due to (a) partitioning of space (e.g. mixture of deep root vs shallow rooted species) (b) separation in time (e.g. different growth phenologies).	Necessary to identify species with complementary traits to ensure stable mixture.
Increased productivity because of greater efficiency in light interception.	Depends on complementary photosynthetic requirements of constituent species.
Increased production because of reduced insect and disease problems. This because susceptible species 'hidden' in space or because mixtures alter microclimate and thereby affect insect behaviour or habitats.	Evidence of reduced insect damage in some agricultural trials but, to date, more equivocal results from tree plantation studies.
Increased production if nitrogen-fixing trees used in mixtures (e.g. *Acacia*).	Significant nitrogen fixing only likely to occur in infertile N-deficient soils.
Increased production caused by faster nutrient turnover since diverse litter layers decompose faster.	
Improved tree form and bole length.	May occur if one species acts as 'trainer' and induces more rapid height growth in others in mix (may also occur in closely spaced monoculture plantations).
Improved profitability if faster growing species (such as eucalypts) are included in mixture and can be harvested early to recoup plantation establishment cost. This thinning promotes growth of the slower growing (more valuable) residual species.	Usually no market for small sized rainforest trees, except for smaller eucalypt poles or posts. Sequential harvesting requires species to be planted in alternate rows to avoid logging damage to the residual trees at time of first harvest.

Sources: Grijpma, 1979; Brown and Ewel, 1987; Binkley, 1992; Simmonds and Buckley, 1992; Kelty, 1992; Wormald, 1992; Lamb and Lawrence, 1993; Keenan *et al.* 1995; Kelty and Cameron, 1995; Montagnini *et al.* 1995.

Lastly, productivity may be enhanced by planting species with complementary crown or root architectures that enable a more efficient use of site resources. This appears to be the cause of an early increase in overall plantation productivity in early results from a replicated trial in North Queensland involving *Flindersia brayleyana* grown in mixture with species including *Eucalyptus pellita* and *Eleaocarpus grandis* compared with monocultures of these species (Bristow *et al.*, in press).

These studies all suggest plantation mixtures may offer production benefits as well as 'conservation' advantages under some circumstances. Some tentative conclusions for mixed-species plantations in North Queensland are outlined in Table 3.3. But such production gains require that the species involved are carefully selected and complementary and that inter-tree competition is less than in monocultures. More work is needed to develop protocols for identifying these complementary species. Plantations containing two or more species will almost certainly be more difficult to manage than traditional monocultures and the biodiversity gains from these simple mixtures will be relatively modest. Perhaps landscape biodiversity gains could be achieved more easily in other ways!

Table 3.3 Some tentative conclusions about species mixtures in North Queensland based on current evidence

1. Simple mixes (only two species) are likely to be more manageable than more complex mixtures involving more than two species. However, mixtures will always be more difficult to manage than monocultures.
2. There is evidence that some species grow better in mixed plantations than in monocultures (e.g. Queensland maple, red cedar) but there may be some species that grow more poorly. Whether this means a mixture is more financially valuable than a monoculture will depend on stocking rates, the proportions of each species used and the relative price of each species.
3. Insect damage can be substantially reduced by underplanting red cedar beneath a previously established nurse crop. There is little other evidence to date that other types of mixtures used have any benefits in terms of insect damage or disease.
4. There appear to be important benefits of including a nitrogen-fixing species on nitrogen limited sites
5. There appear to be some tree form advantages (e.g. reduced branchiness) in some mixtures that could be a particular advantage in plantations established using unimproved seed collected from natural forests, i.e. where a species has not undergone a selection and breeding programme.

3.5. ALTERNATIVE OPTIONS FOR RE-ESTABLISHING LANDSCAPE BIODIVERSITY

Mixed species plantations are an attractive option to achieve conservation benefits as well as some timber production. However, they should not be seen as the only way in which production and conservation objectives might be achieved. Depending on the design, even plantation monocultures can contribute to regional diversity. Other options are as follows:

Using buffer strips around species monocultures. Species-rich buffers could consist of residual forest or forest restored by planting using a wide diversity of species that are not intended for harvest. Besides conferring biodiversity to a landscape and a degree of connectivity to enhance wildlife movement, buffers can be beneficial as streamside protection and as firebreaks. This approach has been used, for example, in many of the plantations established by the forest service in Queensland.

Using small patchworks of different species monocultures scattered across a landscape. Each small patch would be a traditional plantation monoculture but a mosaic of patches could confer a high degree of heterogeneity and biodiversity to the landscape. By locating particular species on sites within the plantation where they perform best it should also be possible to improve the overall productivity of the plantation system. The general idea is to avoid using a single species in a heterogenous landscape where not all parts of the landscape are equally suitable for that one species. This approach necessarily requires specific knowledge of site-species relationships. Studies of the performance to date of the various species in the CRRP plantings in North Queensland suggest some grow best in lowland situations and others prefer tableland sites (Bristow, 1996). More time will be needed for these to grow further allowing more detailed relationships with site factors such as climate, soil and topography to be evaluated.

Embedding a patchwork of species in a matrix of buffer strips. This is essentially a combination of the first two options above.

Fostering the development of species-rich understories beneath plantation monocultures. It has been observed in a variety of tropical landscapes that many single species plantations can develop an understorey of tree seedlings (Lugo 1992, Parrotta and Turnbull, 1997). This has also occurred in many older plantations in North Queensland (Keenan *et al.*, 1997). At one site a plantation of *Flindersia brayleyana* had more than 120 rainforest tree species present in the understorey. A large variety of other life forms were also present. The extent to which this colonisation occurs depends on the distance of the plantation from a seed source in intact natural forests, the size and number of species these intact areas contain, and the availability of dispersers such as birds or bats able to carry the seed across the intervening landscape. Management can also influence the rate at which these understories develop by regulating plantation size, the plantation species used, the thinning regimes applied and so on.

The extent to which the understorey remains intact and provides a habitat for wildlife depends on the way the plantation is managed. There appear to be four possibilities a manager might consider:

- harvest the plantation as planned and destroy any biodiversity in that harvested area. In this case planned financial gains are realised from the plantation and the biodiversity gains can be seen as a temporary bonus which may be regained again when the plantation is re-established. (This assumes, of course, that only a small area of plantation is felled at one time and that opportunities remain for recolonisation from intact plant and animal communities.)

- abandon the plantation as a production area and transfer it to the conservation estate. This option recognises that any production achieved is of lower social benefit than the conservation gains achieved over the lifetime of the plantation. This is obviously a major change in management direction but social goals can change over the 40-year period of a plantation rotation and this choice may be appropriate in particular circumstances.

- carefully remove sufficient of the plantation trees to pay the plantation establishment debt while leaving as much as possible of the accumulated biodiversity intact. Subsequently, manage the forest as a restoration project for biodiversity purposes. Again, this represents a major change in direction but has the advantage that the transition pays its own way.

- harvest both plantation trees plus any commercially attractive species that have colonised the site (present now as canopy trees or as understorey species). That is, recognise the fact that the plantation monoculture has disappeared and been replaced by an entirely new range of species.

Not all of these options will always be available and the market opportunities and landscape context will affect the nature of the choices available (Keenan *et al.*, 1997; Lamb, 1998).

3.6 PLANTATION ESTABLISHMENT AND MANAGEMENT

The techniques of site preparation and weed control used for eucalypts or exotic pines are generally applicable to most other species. However, the growth of any plantation after the initial seedling establishment stage depends on nutritional requirements being met and on the appropriateness of any silvicultural treatments such as thinning. It is commonly observed that soil fertility declines following forest clearing in the tropics. This suggests planted rainforest species may respond to fertilisers. Most of the sites in North Queensland at which trial plantings have been established to date have been old pastures but these have covered a range of soils and management histories. Pot trials have identified potential nutrient deficiencies in a number of soil types (Webb *et al.*, 1997) but experimental field responses to addition of fertilisers have been variable (Keenan *et al.*, 1998b). Many rainforest species have slow early growth and nutrient demand in the first few years after planting may be low. For example, hoop pine initially grows relatively slowly but accelerates after a few years. It does not respond to fertilisation at planting but can respond after about five years on particular sites (Xu *et al.*, 1995). The same may be true of other rainforest species and responses to fertilisation might therefore be more pronounced once full canopy development occurs. Further studies will be necessary to determine whether this is true.

Subsequent management of the stand depends on the growth rates of the particular species and the desired end product. Plantations grown for pulpwood or firewood (i.e. maximum biomass per hectare) are generally grown on short rotations and clearfelled after 10 to 15 years then regenerated by coppicing or replanting. High-value timber trees are generally grown for sawlogs or veneer (i.e. large individual trees) and are usually grown on longer rotations of 40 years or more. Selective fellings (i.e. thinnings) can be undertaken prior to this time to harvest timber that might otherwise be lost due to competition-induced mortality. Thinnings also provide growing space for the fastest growing individuals with the best stem form allowing them to reach the minimum merchantable size in the shortest period of time.

Silvicultural treatments such as thinning operations can be costly and the smaller sized trees removed are often difficult to market. As a result, forest managers often favour a small number of heavy thinning operations. For example, current prescriptions for *Pinus caribaea* in North Queensland involve a single thinning at age 18-20 years and clearfelling at age 25 years. The slower growing but more valuable *Araucaria cunninghamii* is grown on a 45-50 year rotation and has two or three thinnings over this period. There is no knowledge of the silvicultural requirements of any other of the North Queensland rainforest species with the single exception of *Flindersia brayleyana.* Some historical spacing and thinning studies have been used to develop tentative density management regimes for a monoculture of this species (Keenan *et al.*, 1998a; Keenan *et al.*, 1999). The most appropriate regime depends on the rate at which log value increases with size. If small-sized logs can be sold then several thinnings might be applied during a rotation, e.g. from an initial stocking of 1000sph to 500sph then 250sph and then 125sph. Otherwise, only single heavy thinning to 250sph that drastically opens the canopy and reduces between-crown competition might be adopted. Current growth models, assuming a minimum merchantable diameter of 40cm, suggest that a volume of 410m^3/ha at age 60 could be produced.

All these prescriptions apply to monoculture plantations and no studies have yet been carried out to examine how stand development might be managed in mixed-species plantations. Where the mixture contains complementary species growing at similar rates, a similar prescription to that developed for *Flindersia brayleyana* will probably be appropriate. That is, one or more heavy thinnings to reduce crown competition, depending on marketability of the necessarily small-sized trees removed in the operation. A different schedule will be necessary when the trees in the mixture are growing at different rates. For example, a mixture might contain alternative rows of eucalypts that produces large logs in a short time period planted with a slower growing but more valuable cabinet wood species. Early removal of the rows of fast-growing trees can be achieved with minimal damage to the remaining trees. This is equivalent to a thinning operation, and leaves the slower growing but more valuable cabinet wood species to grow further under conditions of reduced competition. But, at the same time, it targets an established market (e.g. for telephone poles) and provides an early financial return rather than trusting that a market can be found when the time for thinning cabinet woods occurs.

The objective of thinnings is to reduce competition by opening the canopy and allow more light to reach tree crowns and to reduce the demand on soil resources. But heavy thinnings also have the effect of producing conditions favouring further understorey development. This suggests production and biodiversity are both enhanced by thinning. The other possibility is that dense understorey growth may eventually begin to compete with the commercial trees. The timing and circumstances under which this might occur needs more careful monitoring in the field.

3.7. CONCLUSIONS

It will be difficult to develop a timber industry based on plantations of native rainforest species in North Queensland. In this early stage it is usually unclear what levels of production can be achieved on the sites available and many silvicultural issues remain unresolved. These include the most appropriate species to use on particular sites, whether mixtures can be used, the ways these new stands are to be managed and the pests and diseases likely to affect growth. Some of these will be resolved as current plantations and trials mature and some will need further research.

Industry development using rainforest species is also complicated when plantation development is undertaken in small plots and stands, as would be the case in many developing countries. Hence most plantation management will be undertaken by landowners with relatively little silvicultural skill. There is likely to be only limited government finance for research and development into more complex silvicultural systems and, indeed, for other essential elements of a plantation program such as tree improvement and studies of timber quality and utilisation of fast grown trees. This contrasts with the massive government investment in plantation development in temperate Australia, New Zealand and Europe where many years of government-sponsored research and development have effectively removed most of the risks for private growers. This has helped pave the way for many private companies to enter the tree growing business.

Perhaps the single most difficult silvicultural problem to resolve in these new forests is how to balance commercial imperatives and the desire by many landowners for a concomitant improvement in biodiversity, conservation and output of non-wood products. Can plantations of rainforest species be designed to include some of the biological complexity of the original forest and still be economically viable? Early research results suggest some reasons to be optimistic but trade-offs will be necessary and these trade-offs will be economic as well as biological. Little is yet known about the nature or extent of these trade-offs.

REFERENCES

Binkley, D. (1992), 'Mixtures of nitrogen2-fixing and non-nitrogen2-fixing tree species', *Special Publications of the British Ecological Society* (11), Oxford: Blackwell Scientific Publications, pp. 97–123.

Binkley, D. and Ryan, M.G. (1998), 'Net primary production and nutrient cycling in replicated stands of *Eucalyptus saligna* and *Albizia falcataria'*, *Forest Ecology and Management* **112**(1–2), pp. 79–85.

Bristow, M. (1996), 'Species-site relationships: early growth of plantation trees under varying climatic conditions in North Queensland', *Botany*. Brisbane: University of Queensland.

Bristow, M., Lamb, D., Huynh, D.N. and Keenan R.J. (in press), 'Making Plantations Attractive to North Queensland Small-Scale Forest Growers: Early Research Results with Mixed Species Plantations', in Proceedings of *IUFRO 3.08 Small-Scale Forestry International Symposium*, Kuranda, Jan 9–15, 2000.

Brown, B.J. and Ewel, J.J. (1987), 'Herbivory in complex and simple tropical successional ecosystems', *Ecology* **68**(1), pp. 108–116.

Cameron, D. M. and Jermyn, D. (1991), *Review of plantation performance of high value rainforest species*, St. Lucia: CSIRO, Division of Forestry, Queensland.

Dale, J.A. and Johnson, G.T. (1991), 'Hoop Pine forest management in Queensland', in F. H. McKinnell, E. R. Hopkins, and J. E. D. Fox, (eds), *Forest Management in Australia*, Chipping Norton, Australia: Surrey, Beattie and Sons, pp. 214–227.

Evans, J. (1992), *Plantation Forestry in the Tropics*, 2nd edn., Oxford: Clarendon Press.

Francis, P. J. and Shea, G. M. (1991), 'Management of tropical and subtropical pines in Queensland', in F. H. McKinnell, E. R. Hopkins, and J. E. D. Fox, (eds), *Forest Management in Australia*, Chipping Norton, Australia: Surrey, Beattie and Sons, pp. 203–213.

Grijpma, P. (1979), 'Resistance of Meliaceae against the shoot borer Hypsipyla with particular reference to Toona ciliata (M.J. Roem.) *var. australis* (F.V.Muell.) C.D.C.', in J. Burley and B.T. Styles. (eds), *Tropical trees, Variation, Breeding and Conservation*, Linnean Society Symposium, Series No. 2, pp. 69–79.

Keenan, R., Lamb, D. and Sexton, G. (1995), 'Experience with mixed species rainforest plantations in North Queensland', *Commonwealth Forestry Review*, **74**(4), pp. 315–321.

Keenan, R.J., Doley, D. and Lamb, D. (1998a), 'Stand density management in Queensland maple plantations', *Second Management and Growing of Trees Training Conference*, Kooralbyn, Australia.

Keenan, R. J., Hambleton, A., Robson, K. and Webb, M. (1998b), 'Growth response of three rainforest cabinet timber species to fertiliser application in plantations in North Queensland', in A. Schulte and D. Ruhiyat, (eds), *Soils of Tropical Forest Ecosystems: Characteristics, Ecology and Management*, Berlin: Springer, pp. 107–114.

Keenan, R. J., Lamb, D., Woldring, O., Irvine, A. and Jensen, R. (1997), 'Restoration of plant biodiversity beneath tropical tree plantations in Northern Australia', *Forest Ecology and Management*, **99**(1-2), pp. 117–131.

Keenan, R. J., Sexton, G. and Lamb, D. (1999), 'Thinning studies in plantation-grown Queensland maple (*Flindersia brayleyana* F. Muell.) in north east Queensland', *International Forestry Review*, **1**, pp. 70–78.

Kelty, M. J. (1992), 'Comparative productivity of monocultures and mixed species stands', in M. J. Kelty, B. C. Larson and C. D. Oliver, (eds), *The Ecology and Silviculture of Mixed Species Forests*, Dordrecht: Kluwer Academic.

Kelty, M. J. and Cameron, I. R. (1995), 'Plot designs for the analysis of species interactions in mixed stands', *Commonwealth Forestry Review*, **74**, pp. 322–332.

Lamb, D. (1998), 'Large-scale restoration of degraded tropical lands: the role of timber plantations', *Restoration Ecology*, **6**(3), pp. 271–279.

Lamb, D. and Lawrence, P. (1993), 'Mixed species plantings using high-value rainforest trees in Australia', in H. Lieth and M. Lohman, (eds), *Restoration of Tropical Rainforest Ecosystems*, Dordrecht: Kluwer Academic Publishers, pp. 101–108.

Lewty, M. and Last, I. (1998), 'Hoop pine plantation management in Queensland', in R. Dyason, L. Dyason and R. Garsden, (eds), *Plantation and Regrowth Forestry: A Diversity of Opportunity*, Australian Forest Growers, Lismore, NSW, pp. 189–198.

Lugo, A. E. (1992), 'Tree plantations for rehabilitating damaged forest lands in the tropics', in M. K. Wali, (ed.), *Ecosystem Rehabilitation*, The Hague: SPB Academic Publishing, pp. 247–255.

Montagnini, F., Gonzalez, E., Porras, C.and Rheingans, R. (1995), 'Mixed and pure forest plantations in the humid neotropics: a comparison of early growth, pest damage and establishment costs', *Commonwealth Forestry Review*, **74**, pp. 306–314.

Nikles, D. G. and Lee, D. (1998), 'Improved varieties of forest trees available or being developed for commercial planting in Queensland and northern NSW', in R. Dyason, L. Dyason and R. Garsden, (eds), *Plantation and Regrowth Forestry: A Diversity of Opportunity*,. Australian Forest Growers, Lismore, NSW, pp. 139–154.

Palmer, J. R. (1992), 'Designing commercially promising tropical timber species', in R. R. B. Leakey and A. C. Newton, (eds), ITE Symposium no. 29 / ECTF Symposium No. 1: *Tropical Trees: the Potential for Domestication and the Rebuilding of Forest Resources*, Heriot-Watt University, Edinburgh, pp. 16–24.

Parrotta, J. A. and Turnbull, J. W. (1997), 'Catalyzing native forest regeneration on degraded tropical lands', *Forest Ecology and Management*, **99**(1–2), pp. 1–8.

Simmonds, E. and Buckley, P. (1992), 'The ground vegetation under planted mixtures of trees', *Special Publications of the British Ecological Society* (11), Oxford: Blackwell Scientific Publications, pp. 211–231.

Vanclay, J. K. (1994), 'Sustainable timber harvesting: simulation studies in the tropical rainforests of North Queensland', *Forest Ecology and Management*, **69**(1–3), pp. 299–320.

Wadsworth, F.W. (1998), *Forest production for Tropical America*, USDA Forest Service Agricultural Handbook 710, Washington DC: USDA Forest Service, Government Printing Office.

Webb, M.J., Reddell, P., Hambleton, A. and Mazza, G. M. (1997), 'Nutritional constraints to growth of Australian red cedar *(Toona ciliata)* in five North Queensland soils', *Australian Forestry*, **60**, pp. 46–62.

Wormald, T. J. (1992*), Mixed and Pure Forest Plantations in the Tropics and Subtropics*, Rome: FAO.

Xu, Z. H., Simpson, J. A.. Osborne, D. O and Bubb, K. A. (1995), 'The role of fertilisation in improving the productivity of hoop pine plantations', Hoop Pine Establishment Workshop, Gympie.

TIMBER BENEFITS AND MARKET CONSIDERATIONS

4. Financial Analysis of a Two-Species Farm Forestry Mixed Stand

J.L. Herbohn and S.R. Harrison

In recent years, there has been considerable interest in growing mixtures of native timber species to develop plantations which have increased biodiversity value. However, typically little information is available about the likely financial performance of these non-traditional forestry systems. This chapter discusses the financial analysis of farm forestry in the Wet Tropics of Australia, from the perspective of a two-species mixture of *Flindersia brayleyana* and *Eucalyptus cloeziana*. A financial model for this mixture has been developed using data collected by a Delphi survey approach[1]. Financial analyses suggests this mixture will generate a positive net present value at a discount rate of 7% provided stumpage prices of between \$150 to \$200/m^3 can be obtained for *F. brayleyana*. Financial modelling is found to present considerable difficulty in terms of scheduling successive harvests and making yield and price estimates. Many of the principles of financial analysis, while discussed in the context of a specific application for North Queensland, have general applicability to other tropical areas.

4.1 INTRODUCTION

There has been great interest by private landholders in growing native tree species on their land in North Queensland. Many native timber species mixtures are available and there is a dearth of information about harvest scheduling and economic performance. A wide variety of species mixtures and harvest systems could be adopted in small-scale forestry using native rainforest and eucalypt species (as discussed in Chapter 3).

1. The assistance of the North Queensland Joint Afforestation Board and the Rainforest CRC in funding the research reported here is gratefully acknowledged.

As part of their research within the Cooperative Research Centre for Tropical Rainforest Ecology and Management, the authors have developed financial models for mixtures of native species grown in a plantation situation, such as takes place under the Community Rainforest Reforestation Program (CRRP) in north Queensland. Financial modelling has involved efforts to develop a general spreadsheet model, as well as to develop models to predict returns from specific mixtures. This chapter summarises some of the research on this project. A financial model for a relatively simple two-species mixture of *Eucalyptus cloeziana* (Gympie Messmate) and *Flindersia brayleyana* (Queensland Maple) is presented. Estimates of growth rates, harvest timing and expected annual cash flows are derived, based on the data from a Delphi survey. The assumed site is degraded farmland on the moderately wet and fertile Atherton Tableland, although performance of this mixture would probably be quite similar in many coastal areas of Queensland and northern New South Wales. Net present values are derived for alternative harvest regimes. The following section discusses data collection and modelling procedures. Estimates of revenue and overall financial performance and a sensitivity analysis are then presented, followed by some concluding comments.

4.2 SURVEY METHOD AND DEVELOPMENT OF THE FINANCIAL MODEL

A planting and harvesting scenario has been developed for a 50:50 mixture of *Flindersia brayleyana* and *Eucalyptus cloeziana* (Table 4.1). Management of the plantation is designed to maximise timber yield of *F. brayeleyana*. Crown shyness of this species, and the associated 22:1 crown diameter to diameter at breast height (dbh) relationship, are allowed for in this management. Thinning and harvesting regimes are timed to occur just as *F. brayleyana* crowns touch, at which time lock-up of growth would be expected to occur.

Requirements for poles suitable for use by electricity authorities have been obtained from North Queensland Electricity Board (NORQEB). NORQEB usually requires plantation-grown poles to conform to the dimensions of poles with a strength rating of one group above that required for native forest sourced poles, i.e. greater butt and head diameters. *Eucalyptus cloeziana* is classified as an S2 species but due to NORQEB requirements, dimensions used are those for an S3 species (11m bole length, 40cm at 2m from butt, small end diameter 26.5cm).

Table 4.1 Planting and harvesting scenario for a Flindersia brayleyana and Eucalyptus cloeziana mixture

Stage and activity	Density after treatment (stems/ha)	Estimates requested from participants in Delphi survey of reforestation experts
1. Plant alternating rows of maple and eucalyptus at 3m x 5m spacing	660 (330 *F. brayleyana*, 330 *E. cloeziana*)	
2. Thin to waste every second tree., when *F. brayleyana* reaches 18cm dbh.	340 (170 *F. brayleyana*, 170 *E. cloeziana*)	Age at which *F. brayleyana* expected to reach 18cm dbh
3. Thin every second eucalypt to waste or for strainer posts. Thin when *F. brayleyana* reaches a dbh of 32cm.	255 (170 *F. brayleyana*, 85 *E. cloeziana*)	Age when *F. brayleyana* expected to reach 32cm dbh. Bole dbh, small-end diameter, bole length of *E. cloeziana* at this age
4. Remove every second *F. brayleyana* for strainer posts or small diam. logs	170 (85 *F. brayleyana*, 85 *E. cloeziana*)	
5. Remove remaining eucalyptus for poles.	85 *F. brayleyana*	Ages when *E. cloeziana* expected to reach 5 specified pole dimensions
6. Harvest every second *F. brayleyana* when crowns touch (50cm dbh)	43 *F. brayleyana*	Age at which *F. brayleyana* expected to reach 50cm dbh Small-end diameter and bole length at this time
7. Harvest remaining *F. brayleyana* (81cm dbh)	Nil	Age at which *F. brayleyana* expected to reach 81cm dbh Small-end diameter and bole length at this time

Personal interviews were conducted with five North Queensland forestry experts chosen for their familiarity with the species being modelled. At the commencement of interviews the plantation system was outlined and the requirements for information stated. Some discussion of expected growth rates and timing of harvest occurred at the end of the session. After collation of estimates, outliers were identified and clarification was sought from participants.

Establishment and tending costs are based on those of Moore (1992). Pruning costs are based on a quotation for contract pruning (Clarke, 1996), with lifts to 2.7m, 4.6m and 6.5m. It is assumed that thinning of every second tree will occur when *F. brayleyana* reaches a dbh of 18cm, at year 8. Certification of pruning is included to reduce uncertainty associated with timber quality and any negative effects on timber prices, with certification costs based on Clarke (1996).

Prices for poles have been determined by reference to Department of Primary Industries Circular 2241. The stumpage value of thinnings of small diameter *F. brayleyana* and *E. cloeziana* is estimated at \$30/m^3. The base price adopted for *F. brayleyana* is \$300/m^3. This is greater than current stumpage paid for maple logs harvested from native forests but below the estimate of \$340 obtained in a survey of expert opinion about potential performance and prices of rainforest species grown under plantation conditions in north Queensland (Russell *et al.*, 1993). Discounted cash flow analysis is used to assess financial performance under various scenarios and to examine sensitivity of performance to parameters such as discount rate and stumpage value. A break-even analysis has been undertaken.

4.3 ESTIMATES OF REVENUE AND NET PRESENT VALUE AND SENSITIVITY ANALYSIS

Estimates of model parameters provided by the Delphi survey and used in predicting harvest volumes and timing of harvest of the *F. brayleyana*/*E. cloeziana* plantation are summarised in Table 4.2. Averages of parameter estimates provided by experts have been used to determine harvest timing of for *E. cloeziana* poles (Table 4.3) and timing of harvest and harvest volume of *F. brayleyana* and *E. cloeziana* for sawlogs (Table 4.4).

Potential annual cash flows have been estimated for a two-species mixture, as in Table 4.5. The first harvest of *F. brayleyana* is predicted to take place in year 34 and produce revenue of approximately \$25,000/ha, while the second harvest is expected to generate revenue in the order of \$80,000/ha. If all *F. brayleyana* trees are harvested at year 34, revenue of approximately \$51,000/ha would be expected. Clearfelling all *F. brayleyana* at age 34 produces the highest NPV. The discount rate used in the calculation of NPV greatly affects overall profitability.

Table 4.2 Estimates of model parameters for a Flindersia brayleyana and Eucalyptus cloeziana mixed plantation

Stage	Parameter	Estimated values from Delphi survey	
		Average	Range
2.	Age when *F. brayleyana* expected to reach 18cm dbh	8.6 years	7 –10 years
3.	Age when *F. brayleyana* expected to reach 32cm dbh	17.6 years	15 – 20 years
	At this age, the following are expected for *E. cloeziana*		
	- bole dbh	41.4cm	35 – 50cm
	- small end diameter	26.2cm	15 – 35cm
	- bole length	16.4m	12 – 20m
5.	Age when *E. cloeziana* expected to reach specified dimensions for		
	- Pole 1	17.4	10 – 25 years
	- Pole 2	17.6	11 – 25 years
	- Pole 3	21.0	12 – 30 years
	- Pole 4	24.2	14 – 35 years
	- Pole 5	25.6	15 – 35 years
6.	Age when *F. brayleyana* expected to reach 50 cm dbh	34 years	25 – 40 years
	Expected small-end dia. at this age	33 cm	30 – 40 cm
	Expected bole length at this age	14.8 m	6 – 20 m
7.	Age when *F. brayleyana* expected to reach 81cm dbh	60 years	50 – 65 years
	Expected small-end dia. at this age	57cm	45 –60cm
	Expected bole length at this age	16.6m	10 – 20m

Sensitivity analysis has been performed to determine the effect of discount rates in the range 6% to 8 %. If all *F. brayleyana* trees are harvested at age 34, NPVs are positive for rates up to 8%. If harvesting is undertaken in two stages (years 34 and 60), the NPV at a discount rate of 8% is negative.

Sensitivity analyses have been conducted on the effect of changes in stumpage prices for *F. brayleyana* on NPV (Table 4.6). If plantations are primarily established for commercial reasons and are managed accordingly, use of a discount rate of 7% for the making of decisions is justified. For purely commercial plantings, the analysis indicates that stumpage prices of approximately \$200/m^3 would be needed for a positive NPV (at 7%) for a two-stage harvest. If, however, all *F. brayleyana* were harvested at age 34, a stumpage price of approximately \$150/m^3 only would be required to produce a positive NPV. Stumpage prices of greater than \$200/m^3 should be readily achievable given the high regard in which *F. brayleyana* is held. For instance, Russell *et al.* (1993) estimated the potential stumpage value of this species to be US\$238/m^3 (about A\$340/m^3) in 1993 prices.

Table 4.3 Predicted harvest ages and returns from Eucalyptus cloeziana poles

Estimated variable	Pole category				
	Pole 1	Pole 2	Pole 3	Pole 4	Pole 5
Number of poles per ha	85	85	85	85	85
Estimated time of harvest	17	18	21	24	26
Price per pole	49	68	77	106	148
Cash inflow (\$/ha)	4,155	5,807	6,567	9,049	12,544
PV (7%)	1,315	1,718	1,586	1,784	2,160

Table 4.4 Estimated timber production from a Flindersia brayleyana and Eucalyptus cloeziana mixture

Species and harvest	Year of harvest	No. of stems harvested (ha)	Harvest volume (m^3/ ha)
E. cloeziana thinnings	17	85	125
F. brayleyana thinnings	17	85	45
E. cloeziana poles (pole cat. 5)	26	85	n/a
F. brayleyana (harvest 1)	34	42	84
F. brayleyana (harvest 2)	60	43	267

The analysis using a discount rate of 7% clearly indicates the optimal harvest scenario to be a single cut at 34 years as opposed to a two-stage cut (34 years and 60 years). It would be expected that timber grown over a longer period of time (e.g. 60 years opposed to 34 years) would have superior colour, grain and density and would attract premium prices compared with younger and faster grown timber. The extent that this quality would translate into higher stumpage prices is unknown. However even if a 50% premium is attached to timber harvested at 60 years compared with that harvested at 34 years, the optimal scenario still remains the single shorter rotation harvest at 34 years.

Table 4.5 Costs and returns of Flindersia brayleyana and Eucalyptus cloeziana mixture

Cashflow variable	Year when incurred or recd.	Amount ($/ha)	NPV, discount rate of:		
			6%	7%	8%
Expense item					
Establishment	0	(2,955)	-2,955	-2,955	-2,955
Pruning	2, 4, 6	(1,419)	-1,155	-1,118	-1,082
Certification of pruning	2, 4, 6	(660)	-525	-507	-489
Tending	1 – 18	(1,804)	-1,563	-1,499	-1,464
Thinning	8	(400)	-253	-233	-216
Revenue - Scenario 1					
E. cloeziana thinnings ($30/m^3)	17	3,752	1,394	1,188	1,014
F. brayleyana thinnings ($30/m^3)	17	1,354	503	429	366
E. cloeziana poles	26	12,544	2,757	2,160	1,696
F. brayleyana (harvest 1) ($300/m^3)	34	25,224	3,479	2,528	1,843
F. brayleyana (harvest 2) ($300/m^3)	60	80,073	2,427	1,382	791
Revenue – Scenario 2 (*F. brayleyana* harvested at age 34)					
F. brayleyana ($300m^{-3})	34	51,049	7,040	5,116	3,729
Total NPV– Scenario 1			4,059	1,313	(547)
Equivalent annual return ($/ha)			63	22	(9)
Total NPV – Scenario 2			5,193	2,519	549
Equivalent annual return ($/ha)			153	74	16

It could be expected that older plantations – particularly those which involve native species – will provide considerable non-market benefits to the landholder and the wider community, e.g. land and watershed protection, wildlife habitat, increased biodiversity (particularly if an understorey is allowed to establish) and improved landscape aesthetics. The levels of these benefits accrue at an increasing rate as the plantation matures. Where non-market benefits are considered important, preferably non-market valuation techniques would be applied to value them. In the absence of such estimates, sometimes a lower 'social' discount rate (e.g. 4% rather than 7%) is used to reflect the important non-market benefits, particularly the growing of native species. However, since there is often no formal quantification of non-wood benefits, the use of such a social discount rate becomes a highly arbitrary approach. If a discount rate of 4% is used, the optimal harvest scenario becomes a two-stage cut with an initial harvest at 34 years and a final cut at 60 years. The choice of this scenario becomes even more clear-cut if a premium is attached to timber harvested at 60 years.

Table 4.6 Effect of variation in stumpage prices on NPV of two-species plantations

Harvest regime and price premium	NPV for a stumpage price of ($ m^{-3})				
	50	100	150	200	300
7% discount rate					
F. brayleyana - 1 harvest, 34 yrs	-1,744	-891	-39	814	2,519
F. brayleyana - 2 harvests, no age premium on 60 y.o. timber	-1,945	-1,293	-642	10	1,313
F. brayleyana - 2 harvests, 25% age premium on 60 y.o. timber	-1,887	-1,178	-469	240	1,659
F. brayleyana - 2 harvests, 50% age premium on 60 y.o. timber	-1,830	-1,063	-296	471	2,004
4% discount rate					
F. brayleyana - 1 harvest, 34 yrs	2,597	4,840	7,082	9,324	13,809
F. brayleyana - 2 harvests, no age premium on 60 y.o. timber	2,732	5,108	7,485	9,861	14,615
F. brayleyana - 2 harvests, 25% age premium on 60 y.o. timber	3,049	5,743	8,436	11,130	16,518
F. brayleyana - 2 harvests, 50% age premium on 60 y.o. timber	3,366	6,377	9,388	12,399	18,421

It is possible that increasing demands on the world's timber supply will result in real increases in timber prices in the future. For instance, the softwood pulpwood royalty increased at an average rate of 0.68% p.a. in real terms from 1969 to 1996 (Townsend and Mahendrarajah, 1996). Russell *et al.* (1993) suggested an annual real increase of 1.3%, which they considered conservative compared to a number of long-term forecasts.

Sensitivity analyses of real increases in timber prices ranging from 0.5% to 1.3 are reported in Table 4.7.

Table 4.7 Effects of real increases in timber prices on returns from a two-species mixture

Harvest regime	NPV @ 7%, real price increase of:			
	0%	0.5%	1.0%	1.3%
F. brayleyana (2 harvests, 34 yrs and 60 yrs)	1,313	2,405	3,757	4,721
F. brayleyana (1 harvest, 34 years)	2,519	3,607	4,877	5,737

Note: Base stumpage prices for thinnings harvested at 17 years is $30/m^3, and $300/m^3 for *F. brayleyana* harvested at 34 and 60 years.

Real increases in timber prices greatly affect the NPV of the plantation system, both for a clear-fell at 34 years and a partial cut at 34 years and clear-fell at 60 years. An even more dramatic effect is seen on cashflows at the time of harvest. For example, assuming a base price of $300/m^3 and a real price increase of 1.3%, total harvest revenue at year 60 would be about $180,000/ha, compared with $80,000/ha if no increase in real price occurred.

This study constitutes what is probably the first attempt in Australia to model timber production and revenue for a two-species mixture of native timbers. Much uncertainty remains with respect to expected yields, timber prices and harvest ages. Given the limited data available for making such predictions, a Delphi survey such as that used in this study is arguably the only methodology by which reasonable predictions of returns can be made.

4.4 CONCLUDING COMMENTS

Lack of information about profitability of growing native timbers has probably held back development of the industry. Financial models have been developed and are well accepted for monocultures of exotic conifers, and even for the native conifer Hoop Pine. However, development of financial models for mixtures of native species presents considerable difficulty, because of the great variety of species combinations mixtures possible, lack of knowledge of harvest ages and timber yields, high price uncertainty, and a variety of other factors. In defining a particular two-species mixture and associated harvest regime, and estimating growth rates, this study has attempted to shed some light on potential returns.

REFERENCES

Clarke, F.B. (1996), 'Pruned stand certification in Australia', *Farm Forestry and Plantations*, 1996 Australian Forest Growers Conference, Mt Gambier, pp. 232–236.

Moore, P. (1992), *Rainforest Timbers and Cabinetwoods: Marketing Considerations*, paper presented to Mt. Mee forestry field day, Brisbane: Queensland Forest Service.

Russell, J.S., Cameron, D.M., Whan, I.F., Beech, D.F., Prestwidge, D.B. and Rance, S.J. (1993), 'Rainforest trees as a new crop for Australia', *Forest Ecology and Management*, **60**(1): pp. 41–58.

Townsend, P.V. and Mahendrarajah, S. (1996). 'What influences the returns to plantation forestry?', *Farm Forestry and Plantations*, 1996 Australian Forest Growers Conference, Mt Gambier, pp. 160–171.

5. Estimating Investment Risk in Small-Scale Plantations of Rainforest Cabinet Timbers and Eucalypts

S. R. Harrison, J. L. Herbohn and N. F. Emtage

Farm forestry is not a risk-free investment. Yield and price uncertainty arise due to lack of knowledge about stand growth performance, particularly for non-traditional species, and the long lead time from planting to harvest. As well, sovereign risk arises with respect to harvest rights and environmental regulations. If estimates can be made of the probability distributions of yield and prices, then risk analysis can be performed to estimate the probability distribution of financial performance. Such distributions are derived for a number of native species and mixtures in reforestation plantings in North Queensland, revealing a high level of investment risk.

5.1 INTRODUCTION

Farm forestry based on native timber species may be regarded as a form of long-term investment and, therefore, financial performance should be estimated by criteria such as net present value (NPV) and internal rate of return (IRR). Estimates may be derived based on best-bet timber yields, production costs and prices. However, parameter estimates often involve a high level of uncertainty. Literature suggests that investors are, in general, risk averse, and information about risk as well as expected returns is important when comparing forestry with other investments, and even between alternative forestry options. The technique of *risk analysis* or venture analysis provides a useful approach to estimating investment risk. In this technique, probability distributions are attached to the key parameters determining cash flows, and probability distributions of performance criteria such as NPV are generated. This provides the decision-maker with greater information than from a deterministic analysis. Risk analysis may be viewed as an alternative to sensitivity analysis, in which not only the possible ranges of parameter values but also probabilities over these ranges are taken into account.

This chapter first examines concepts of risk and uncertainty. Sources of risk in farm forestry are then reviewed, and attitudes of decision makers to risk discussed. Next, methods of assessing investment risk in farm forestry are considered, and a case study of risk evaluation is presented.

5.2 THE NATURE OF 'RISK' AND 'UNCERTAINTY' IN FARM FORESTRY

When funds are to be invested now which will result in revenue being generated perhaps decades into the future, it is impossible to predict the payoff exactly. Some types of investments – such as purchase of fixed interest bonds – have a guaranteed return, but these are exceptions. Where the return is subject to biological and market factors, there can be a wide distribution of possible financial outcomes; the investment payoff is subject to *risk* or *uncertainty*. Many definitions may be found for these terms. Both terms refer to lack of knowledge about future values of *random variables* relating to financial performance. In a project evaluation context, variables which enter cash flows do not take single-point or deterministic values but rather their future levels can be described by probability distributions, arising from unpredictable factors such as weather, market conditions, changes in technology and legal or institutional factors.

A distinction is sometimes drawn between the two terms. 'Uncertainty' is applied to random variables for which it is not possible to specify the parameters of probability distributions – for example, when there is no history of observations upon which to draw. 'Risk' on the other hand arises when it is possible to derive estimates of the probabilities based on past records e.g. the risk of not receiving a particular amount of rainfall in a given month at a specific location. In practice, most events are *non-repeatable* (even rainfall in the above example may be subject to climate change), so uncertainty is more common than risk. In practice probability distributions of cash-flow variables are usually estimated *subjectively*, based on whatever records are available, often along with experience and judgement. This is sometimes described as converting uncertainty to *quasi-risk*. Nowadays, the terms 'risk' and 'uncertainty' are often used interchangeably, and variability in physical or financial performance is frequently referred to as 'risk'. While it is convenient to think of 'risk' with respect to variability both above and below expected values, the main concern is usually with *downside risk*.

Major sources of variability in financial performance of farm forestry are summarised in Table 5.1. Most of the plantation establishment and maintenance costs are incurred in first few years of plantation life, and are reasonably predictable. Contract rates for operations are generally well known. Cost of land preparation can vary to some extent, depending on site characteristics. Seedling costs should be known in advance, as well as initial fertilizer costs. Weed control practices and labour requirements can be predicted within a narrow range. Should unusually dry weather or damage by pests or animals be experienced after planting, there may be additional costs in watering and replanting lost trees.

Table 5.1 Sources of risk in farm forestry

Risk category	Major sources
Risk of poor establishment	Dry weather, inferior planting stock, poor planting technique, poor weed control
Production (timber yield) risk	Storm or cyclone Fire Pest (insect, disease) Unsuitable species mixture Collateral damage at harvest
Timber quality and product type risk	Inappropriate silvicultural management Insect damage Whether high-value outputs are produced, e.g. cabinet timber, poles and peelers vs. sawlogs
Sovereign risk	Machine use and roading regulations Changes in taxation arrangements Uncertain harvest rights and compensation
Market risk	Uncertain future timber prices

Tree losses and growth rates are difficult to predict, particularly when species are used for which limited experience has been gained. Even with the best of intentions, there is no guarantee that thinning and pruning will be carried out at optimal times. The future prices for different products (e.g. poles, peelers, cabinet timber, house framing timber) that can be produced and the best product mixes are also uncertain. For example, high quality poles are likely to be more profitable than sawlogs due to earlier harvest and favourable prices; however, this assumes the existence of a future market.

With mixed species plantings and differing harvest ages for different species, there is a risk of collateral damage to retained trees at the various harvest stages. This is overcome to some extent by planting single species within rows, but even this system presents problems in that growth within rows will not be uniform and it may be desired to grow some individual trees on. Sovereign risk relates to changes in the rules relating to farm forestry by government. Over the life of a plantation, it is probable that there will be numerous changes in taxation arrangements. Also, environmental regulations could become more stringent, and increase management and harvesting costs or even lead to outlawing of harvest. The extent to which compensation will be provided for such limitations on property rights is unpredictable.

Price risk can arise due to changes in environmental controls (e.g. the 1993 price spike was related to concern over withdrawal of timber resources for conservation associated with the Spotted Owl in the USA), changes in the state of the international economy (as reflected for example in timber prices falls during the Asian financial crisis), development of timber substitutes, and other factors.

In general, the sources of risk outlined in Table 5.1 do not pose a likelihood of total crop loss. Even with severe storm damage or fires, it is usually possible to salvage a considerable proportion of the timber, especially for small plantations. In summary, a variety of sources of risk arise with respect to timber quantity, quality and price. Some of the measures for combating threats to production involve increased cost.

Superimposed upon the various risk components which affect the payoff from farm forestry, individual landholders often do not have access to existing information about plantation profitability, or details of soil nutrient status, pH and so on for the proposed site. This lack of information adds to the uncertainty when developing a budget for a forestry investment, though it may be reduced though access to extension services and planning.

5.3 BUSINESS IMPLICATIONS OF ENTERPRISE RISK

The importance of investment risk will depend on the goals and attitudes of the decision-maker. In general, people are less likely to undertake investments if financial risk is high, and if they do, will require a corresponding higher rate of return. Also, for two projects with the same expected payoff, the one for which the spread of possible outcomes is least tends to be the preferred option. The reason is that most people (investors) are *risk averse*. Of course, some people derive pleasure (utility) from undertaking a gamble, i.e. are *risk preferers*.

Decision-makers' tradeoffs with regard to investment risk may be viewed in terms of the probability distribution of payoff criteria. Suppose two alternative investments have the same expected payoff (E) but one has lower payoff variance (V) as reflected in a more narrow and peaked payoff distribution. The alternative with the lower variance may then be judged preferable for a risk averse investor.

Higher payoffs are usually associated with higher risks, such that investors frequently have to make a choice along their *E-V indifference curve*. The implications for investment decisions are that the rate of payoff must be higher for risky than for safe investments, that is, high financial risk is a disincentive to investment. If the level of risk is within the normal operating range of revenue of a farm business, then it may not be cause for concern. This could be the case where a landholder places only a few hectares under forestry. But for larger areas, where the performance of the investments can influence the reputation and future financial viability of the farm business, explicit measurement of risk may be important.

5.4 METHODS OF INCORPORATING PAYOFF RISK IN DISCOUNTED CASH FLOW ANALYSIS

The technique of discounted cash flow (DCF) analysis is normally used to evaluate farm forestry investment proposals. Most forestry expenditure is incurred in the first few years, with returns from commercial thinnings then final harvest after perhaps 30 years or more. DCF analysis makes allowance for the differences in timing of costs and returns, providing performance criteria such as NPV, land expectation value (LEV) and IRR. A variety of methods have been devised to take account of risk in DCF analyses. These provide additional information so that project accept/reject decisions and choices between alternatives can be made which accord with the attitudes of decision-makers towards risk. The choice depends to some extent on how important explicit recognition of risk is deemed to be, and how critical risk appears to be in terms of its impact on overall project viability.

Taking conservative benefit estimates. This is probably the simplest and most widely adopted method of allowing for project risk. However, benefits tend to be discounted in a rather arbitrary manner, and the approach provides little information to the decision-maker about the extent of risk faced. The emphasis is on protection against 'downside risk', and no recognition is given of the possibility of payoffs above the single-point estimates. As a result, projects which are financially sound could easily be rejected. A variant of this approach would be to take inflated estimates of project costs. A more complex variation (not elaborated upon here) is to derive 'certainty equivalents' for project benefits.

Requiring a short payback period. Sometimes projects are chosen because they lead to recovery of expenditure in a short period (say five to seven years). This risk-averse criterion can be criticised for making use of partial information only about project payoff. In clearfell plantation forestry, the payback period is normally the harvest age, so this criterion provides no additional information to the decision-maker.

Loading the discount rate with a risk margin. Raising the discount rate, and hence reducing the NPV, is a means of obtaining a conservative estimate of a project's payoff. This has a similar effect to using conservative revenue estimates. Also, it biases the evaluation in favour of short rotation forestry.

Sensitivity analysis is the widely adopted method of measuring project risk. If cash flows are entered on a spreadsheet, this involves simply changing the values of selected parameters in the spreadsheet (by say ±30%) and recording the new values of the performance criteria, and finally presenting the results in tabular form. These calculations can be automated by use of spreadsheet functions such as the *Table* function in Excel. Sensitivity analysis is usually carried out with respect to the variables for which changes in values are likely to have the greatest impact on overall project performance, typically revenue items and the discount rate. While sensitivity analysis can be conducted with respect to parameters one at a time, it is more useful to examine the combined impact of changes in values of two or more parameters, and to identify breakeven levels where the project will just pay its way.

Risk or venture analysis is an extension of DCF analysis in which the variability of performance of a project is estimated by fitting probability distributions to the more uncertain cash flow parameters, and subsequently estimating the probability distributions of performance criteria using Monte Carlo sampling. Annual cash flows are then simulated, and DCF performance criteria are derived. The sequence of steps (generating random parameters values from their probability distributions, calculating the cash flow sequence, deriving performance criteria) is repeated a large number of times, yielding a large number of random values of a performance criterion (say NPV), which can be presented as a cumulative relative frequency curve. Relative to sensitivity analysis, this approach has a greater input data requirement but provides more information to the decision-maker.

Stochastic dominance analysis is a relatively complex extension of risk analysis, which allows alternative investments to be compared on the basis of their cumulative relative frequency distributions of payoffs, even when these distributions intersect (e.g. see Anderson *et al.*, 1977).

5.5 APPLICATION OF RISK ANALYSIS TO FARM FORESTRY IN NORTH QUEENSLAND

Risk analysis has been applied to single species and mixed species plantings, using the Australian Cabinet Timbers Financial Model (ACTFM) developed by members of the Cooperative Research Centre for Tropical Rainforest Ecology and Management (Herbohn *et al.*, 1997; Herbohn *et al.*, 1999). Two major risk components affecting financial performance – yield risk and price risk – have been incorporated into the model, using the @RISK add-on in Microsoft Excel (Palisades Corporation 1995). Cumulative relative frequency distributions have been derived for single species and mixed species plantings.

Data used in the Australian Cabinet Timbers Financial Model (ACTFM) were drawn from a Delphi survey of forestry experts working in, or familiar with farm forestry in North Queensland. Estimates were obtained of likely growth rates and harvest ages of tree species with potential for use in cabinet timber plantations. In a Delphi survey, forestry experts were asked to provide estimates of mean annual increment (MAI, in m^3/ha/year), and harvest age (in years) for 32 tree species that have been used in the Community Rainforest Reforestation Program (CRRP). Estimates included best-guess (or *modal*), most *optimistic* and most *pessimistic* values. The biophysical setting for these estimates was basalt-derived soils, and annual rainfall of 1,600 to 2,000mm per year, in the Atherton Tablelands region of North Queensland.

The average of best-guess estimates are used as the default data for the MAIs and harvest ages in the ACTFM. For the risk analysis, only three species which are popular in farm forestry have been chosen, one of which (*Araucaria cunninghamii* or hoop pine) is grown widely in government plantations. Estimated timber yields and harvest ages of these species are presented in Tables 5.2 and 5.3. Price distributions (Table 5.4) are based on Russell *et al.* (1993) and information from the Department of Primary Industries.

The @RISK Simulation Procedure

A variety of forms of probability distributions are available in @RISK; the relatively simple triangular distribution was used in this study. Distributions were defined for the harvest age, MAI and timber price of each species or mixture, in terms of pessimistic, modal and optimistic estimates. The simulation experiment for each species or mixture involved a number of replications (referred to in @RISK as iterations). For each replicate, random *input* values of all parameters are sampled from their probability distributions, and DCF analysis is performed to obtain the NPV as the *output* or response variable. This procedure was repeated 500 times (a sample size of 500) and the cumulative relative frequency distribution of NPV obtained.

The @RISK package allows correlations to be applied to pairs of parameter values. For this analysis, MAI has been negative correlated with harvest age, with a correlation coefficient of -0.7 (implying that more rapid growth rates would result in earlier harvest). For the two-species models, positive correlations of 0.6 have been imposed on MAI levels (implying that better than expected growth rates of one species are associated with better than expected rates for other species). These correlation coefficients are indicative figures only, with no relevant experimental data being available upon which to base them. Positive correlation between inputs will tend to increase variation in the output variable.

For the price distributions, the hoop pine price in 1998 ($\$100/m^3$) has been taken as the modal estimate for this species and also for *Flindersia brayleyana* (Queensland maple), while a $\$75/m^3$ stumpage price has been taken for *Eucalyptus cloeziana* (Gympie messmate). Estimates from Russell *et al.* (1993) have been taken as optimistic prices. Downside ranges approximate the upside ranges relative to the modal prices. While it has not been feasible to include all the risk components of Table 5.1 into the financial model, price risk is included, and yield risk incorporates much of the establishment and production risk. As well, an allowance is made for catastrophic risk, such as cyclone or storm damage or fire. In particular, the ACTFM model has been modified to include a random 5% possibility that the plantation yield would be unexpectedly reduced by 80%. This approach also provides a proxy for sovereign risk such as introduction of severe limits on what plantation area can be harvested. The probability estimates must be regarded as indicative, with no reliable catastrophic risk data being available.

Table 5.2 Estimated timber yields for selected cabinet timber species in North Queensland from a Delphi survey of forestry experts

Tree species	Mean annual increment (m^3/ ha/ yr)				Number of estimates
	Average– best guess	Average– highest	Average– lowest	Standard deviation –best guess	
Araucaria cunninghamii	19.4	21.9	11.9	11.46	10
Eucalyptus cloeziana	17.2	22.7	12.1	4.68	9
Flindersia brayleyana	13.5	21.1	11.5	9.92	6

Table 5.3 Estimated harvest ages for selected cabinet timber species in North Queensland from a Delphi survey of forestry experts

Tree species	Harvest age (years)				Number of estimates
	Average –best guess	Average –shortest	Average –longest	Standard deviation –best guess	
Eucalyptus cloeziana	35.5	22.8	45.5	10.11	11
Flindersia brayleyana	43.0	34.5	67.0	10.85	10
Araucaria cunninghamii	43.8	33.8	56.5	4.83	12

Table 5.4 Price distributions used in the risk analysis

Species	Pessimistic price ($/m^3)	Modal price ($/m^3)	Optimistic price ($/m^3)
Eucalyptus cloeziana	50	75	100
Flindersia brayleyana	50	100	145
Araucaria cunninghamii	80	100	120

Discount Rates and Price Trends

Estimated financial performance of forestry is strongly dependent on choice of discount rate, and a wide range of rates are to be found in the literature. A real rate (net of inflation) of 7% has been recommended by the Department of Finance (1991) for public sector projects. For various reasons, lower rates are typically used for forestry investments. Reducing the discount rate is an indirect way of allowing for a rate of change in benefit parameters above or below that of the inflation rate. For instance, there may be an increase (or decrease) in real timber prices over time, e.g. an increase of 1.3% is suggested by Russell *et al.* (1993). The discount rate is sometimes reduced to make allowance for non-wood benefits of plantations. It may be possible to produce alternative and more profitable products than those assumed in the financial model – e.g. poles or peelers instead of sawlogs – and reducing the discount rate indirectly allows for this possibility. A further reason may be to allow for taxation benefits, e.g. when planting takes place at a time of high income with harvest after retirement. Discount rates of 4% and 6% have been adopted here.

Risk Analysis for Single Species Plantations

Risk simulation has been carried out for plantations of various individual species. Computer output for hoop pine plantations is presented in Figures

5.1 and 5.2. At the lower discount rate, the probability of a negative NPV is less than 0.2. The possible NPV ranges from a loss of $10,000/ha to a positive payoff of $14,500/ha. At a 6% discount rate, the probability of a loss increases to about 0.8. Also, the cumulative relative frequency curve becomes more vertical, indicating a lower dispersion of payoff levels.

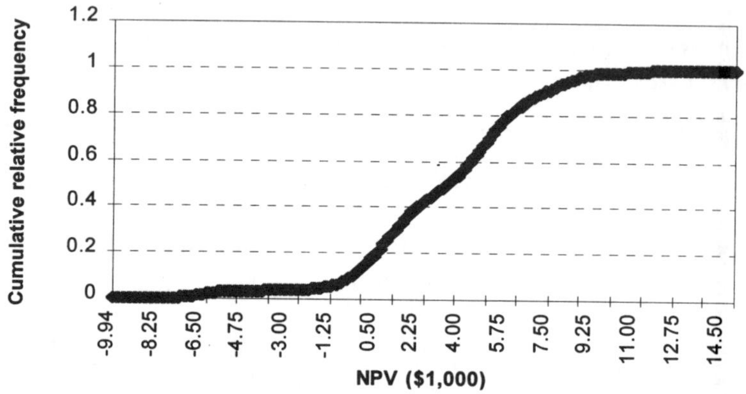

Figure 5.1 Probability of various NPV levels per hectare for a plantation of Araucaria cunninghamii at 4% discount rate

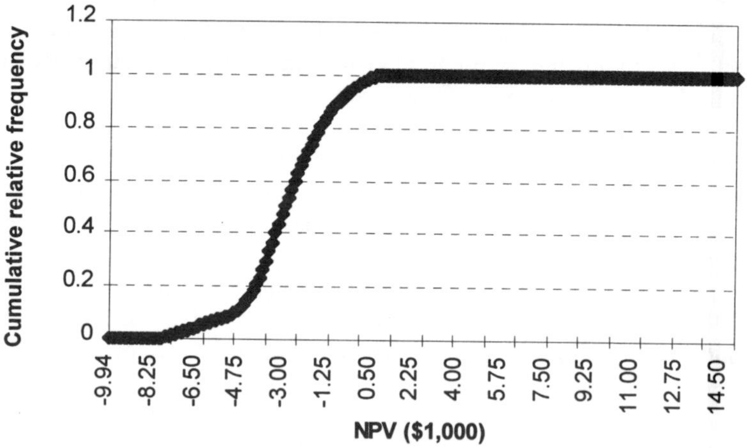

Figure 5.2 Probability of various NPV levels per hectare for a plantation of Araucaria cunninghamii at 6% discount rate

Risk analysis for mixed species plantings

Planting mixtures rather than single species has the potential to reduce risk associated with both yield and price. At present there is little information on which to model relationships between species, and the positive correlation referred to above is a rather simple step towards recognizing this relationship. Figure 5.3 presents output of risk simulation for a mixture of *Eucalyptus cloeziana* and *Flindersia brayleyana*, for a 4% discount rate. There is a relatively high dispersion of NPV levels, with a low probability that the NPV will be negative.

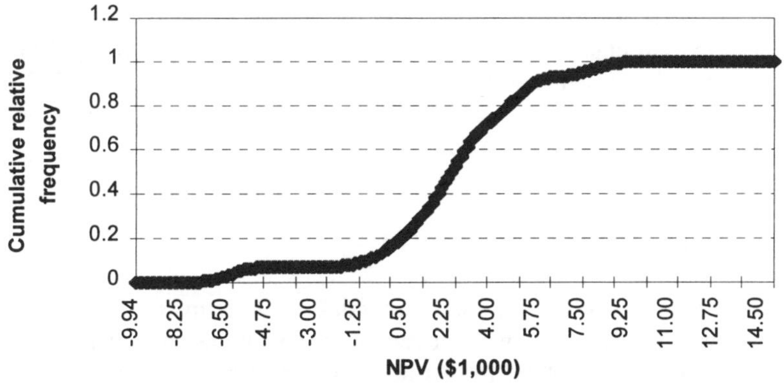

Figure 5.3 Probability of various NPV levels per hectare for a plantation of Eucalyptus cloeziana and Flindersia brayleyana at 4% discount rate

5.6 MANAGEMENT AND POLICY IMPLICATIONS

While a cumulative distribution of performance levels is not as simple to interpret as a single payoff figure, it provides more useful information on which to base decisions. The Delphi survey of forestry experts indicated that a great part of the variation in the timber volumes produced from plantations is attributable to the way in which those plantations are managed (Herbohn *et al.*, 1999). A range of financial returns may serve to stimulate people to seek out the best possible information about how to plan, establish and manage their plantations bearing in mind the differences in potential financial returns from well managed plantations as opposed to those that are poorly managed.

To the extent that high risk is an impediment to farm forestry, as an activity which generates positive off-site benefits for the community, there is a case for government intervention to reduce investment risk in forestry. An obvious avenue is reduction of sovereign risk, since this is under the control of government. More clearly defined environmental regulation and taxation arrangements, and guaranteed harvest rights or rights to compensation where harvesting is not permitted, would be positive steps in this direction.

From a landholder's perspective, there is a need to take account of the not insignificant risk in farm forestry relative to other landuse options. Also, plantation establishment and management measures can be arranged to reduce yield risk. Plantation insurance, forward selling and other options may be possible. Further, mixed species plantations would appear to have potential to reduce payoff risk. Plantation joint venture schemes, which are now common in Australia, go some way towards reducing financial risk in forestry. In particular, they provide more assured quality of plantation management, increased market power to achieve high prices, sharing of risks by the joint venture partners, and reduced risk that property rights will be affected by say new local government bylaws.

5.7 CONCLUDING COMMENTS

Small-scale farm forestry imposes substantial risk in terms of the return on investment. This contrasts with government plantations in which a large forestry investment portfolio results in spreading of risk between sites. On the other hand, many farm forestry ventures are small relative to the total business operation, and hence may not jeopardize the financial soundness of the overall business.

Financial models such as the ACTFM provide an indication of the potential returns from cabinet timber plantations. Compared with use of financial models to calculate the returns for a single scenario, the main difference and advantage of risk analyses is that those interested in establishing plantations are provided with an indication of the range of returns that may result from a plantation and the likelihood that the payoff criterion will be within any particular interval. This contrasts with a single figure based on best estimates, or ranges without probability weights generated by sensitivity analysis.

It is not possible to incorporate all the sources of risk in farm forestry in a risk analysis. However, the most critical ones – relating to yields and prices and catastrophic events – can be simulated. The best source of input data at present for risk analysis on plantations based on native cabinet timbers is subjective estimates by foresters and botanists. Little information is available to model the risk of catastrophic events or the correlation between simulation input variables. Information generated by risk analysis provides a more comprehensive evaluation of farm forestry, as a basis for comparison with other land uses and off-farm investment opportunities.

REFERENCES

Anderson, J.R., Dillon, J.L. and Hardaker, J.B. (1977), *Agricultural Decision Analysis*, Ames: Iowa State University Press.

Department of Finance (1991), *Handbook of Cost-Benefit Analysis*, Canberra: Australian Government Publishing Service.

Herbohn, J.L., Harrison, S.R. and Emtage, N. (1997), 'Financial models for small scale farm forestry: an Australian case study involving mixed species plantations', in Y. Murashima (ed.) *Sustainable Management of Small Scale Forestry*, Proceedings of IUFRO Symposium, the Graduate School of Forestry, Kyoto University, Kyoto: Division of Natural Resource Economics, Kyoto University, pp. 48–54.

Herbohn, J.L., Harrison, S.R. and Emtage, N. (1999), 'Potential performance of rainforest and eucalypt cabinet timber species in plantations in North Queensland', *Australian Forestry*, **62**(1), pp. 79–87.

Palisades Corporation (1995), *@RISK: Advanced Risk Analysis for Spreadsheets*, New York: Newfield.

Russell, J.S., Cameron, D.M., Whan, I.F., Beech, D.F., Prestwidge, D.B. and Rance, S.J. (1993), 'Rainforest trees as a new crop for Australia', *Forest Ecology and Management,* **60**(1), pp. 41–58.

6. Predicting Yields and Returns to Smallholder Plantation Forestry: a Philippines Case Study

T.J. Venn, S.R. Harrison and J.L. Herbohn

Australian tree species are grown widely in tropical and sub-tropical countries throughout the world, and generally exhibit rapid growth rates. Site-species matching and nutrition trials for Australian eucalypts, acacias, casuarinas and *Grevillea robusta* have been carried out recently in the Philippines. Preliminary estimates have been made of yields and returns for these species in smallholder forestry systems[1]. A method is developed to estimate stand yield based on sparse data from the Philippines and elsewhere. Estimates suggest that the Australian species generally will not outperform traditional species in the Philippines on high quality sites, but have a role on poorer sites and for particular end uses.

6.1 INTRODUCTION

A lack of information about yield and returns may limit investment in small-scale plantation forestry to below socially optimal levels. Estimates of yields and returns to plantation forestry are critical for operational and strategic planning purposes. At the operational level, estimates of growth and yield can assist decision-makers to determine optimum planting densities, and guide the timing and intensity of silviculture. The integration of yield and financial models can facilitate strategic decision-making by evaluating the profitability of forestry investments, including comparisons between species (Clutter *et al.*, 1983; Vanclay, 1994).

Establishment of plantation forests in the Philippines began in the 1960s. Despite the official reforestation of 1.5 M ha with exotic species such as *Gmelina arborea* (gmelina), *Paraserianthes falcataria* (falcata), *Tectona*

[1] The research reported here was carried out under an add-on project to ACIAR project SST 96/110, *Tree Production Technologies for the Philippines and Tropical Australia*. Financial assistance from ACIAR is gratefully acknowledged.

grandis (teak) and *Swietenia macrophylla* (mahogany), plantations supply only small volumes of timber (Sy, 1998). During the 1990s, the Australian Centre for International Agricultural Research (ACIAR) funded research into growing plantations of 'Australian timber species' in the Philippines (including *Eucalyptus urophylla* which is in fact endemic to Indonesia). Many of the Australian *Eucalyptus*, *Acacia*, *Casusuarina* and *Grevillea* species were found to grow well under Philippine conditions, even on poor sites. Scientific and anecdotal evidence suggests that several Australian species grow faster, have greater tolerance of poor soils, have a greater ability to recover after fire, and produce denser and more valuable timber than some of the traditional plantation species. For comparative purposes, traditional species were also grown in the ACIAR trials; however, lack of planting stock excluded *G. arborea* and *P. falcataria*. Some trials did include *Eucalyptus deglupta* (bagras), *S. macrophylla*, *Samanea saman* (raintree), *Pterocarpus indicus* (narra) and *Leucaeana leucocephala*, which were all outperformed by Australian species (Venn and Harrison, 2001).

This chapter discusses the development of preliminary models of stand yield and financial performance for Australian tree species on smallholdings in the Philippines, and in particular *Acacia mangium*, *Eucalyptus camaldulensis*, *E. tereticornis*, *E. pellita* and *E. urophylla*. The collection of smallholder yield and financial data is described, and the development of yield and financial models for smallholder plantings in the Philippines is reported. Yield and financial performance estimates of smallholder plantings of Australian and traditionally grown species are compared.

6.2 SMALLHOLDER VERSUS INDUSTRIAL YIELDS AND RETURNS

Many growth and financial models have been developed for various plantation species in the Philippines, though few specifically for small-scale forestry. Often little consideration is given to differences in yield and profitability between industrial and small-scale plantings, and estimates of industrial returns are used in forestry extension, despite scientific and anecdotal evidence suggesting large differences. Particularly in less developed countries, there is a considerable difference between crop yields achieved by farmers and those possible using the best available practices and the actual yields achieved by farmers. For example, rice yields from farms of average productivity in Brazil and Pakistan were approximately one-third of the yields achieved in agricultural research trials (Arnon, 1989). Farmers generally operate under less optimal conditions of site quality and management than experiment stations. Smallholders often have more limited resources and less tenure security than researchers and industrial foresters, which often results in low intensity silviculture. Observations in the Philippines suggest that farmers adopt low intensity management regimes with regard to thinning, pruning and fertilising, leading to trees with poor form, health and vigour. There is also much anecdotal evidence that farm yields are lower and comprise smaller logs than harvests from industrial plantations. Low intensity management may be rational for farmers with few resources, commitments to manage cash crops and a need to meet multiple objectives from their land. Low intensity management may also be the result of poor understanding of silviculture requirements, e.g. the misunderstanding that more trees equates with greater yields and returns.

There is little consensus on whether smallholder or industrial management regimes are more profitable. Niskanen (1998b) found small-scale plantations more profitable than industrial plantations of *E. camaldulensis* in Thailand, due to savings in variable costs (mostly labour). On the other hand, Niskanen and Saastamoinen (1996) concluded that industrial plantations of *S. macrophylla* were more profitable than smallholder plantations in the Philippines. Variability in financial performance is the result of complex interactions between many factors, including yield, cost and supply of labour, and level of influence or control over markets, prices and transport (Venn *et al.*, 2001a).

6.3 COLLECTING SMALLHOLDER YIELD AND FINANCIAL DATA

For species that have only recently been grown in a region, it is unlikely that useful local yield and financial information will be available for modelling, especially for smallholder plantations. In the Philippines, the oldest plantings of eucalypts for which growth data are available were established in 1993, with measurements available for up to five years. While there is a longer history of growing *A. mangium* in the Philippines, little information was available for this study. There are no official data describing markets and prices for Australian timber species grown in the Philippines.

Visits were made to the Philippines to collect smallholder management, yield and financial information for Australian and traditional species. Discussions were held with staff of the Department of Environment and Natural Resources (DENR), Bukidnon Forests Incorporated (BFI), the Philippines Council for Agriculture, Forestry and Natural Resources Research and Development (PCARRD), and with researchers in the Forestry Department and the Forest Products Research and Development Institute (FPRDI) at the University of the Philippines, Los Baños (UPLB). Diameter and height measurements were made by the authors in smallholder timber plantations of *A. mangium* in Bukidnon province, Mindanao and near Matalom on Leyte. Some data on industrial *A. mangium* plantations were obtained from Vendiola (1996). BFI provided expert opinion on industrial yields of *A. mangium* and Australian eucalypts in Bukidnon province. International literature was searched for supplementary growth and yield data for *A. mangium* and Australian eucalypts. While interest has burgeoned recently in growing Australian species in Australia (FWPRDC, 1998) and around the world (Eldridge *et al.*, 1994), knowledge about many tropical Australian species, including those being grown in the Philippines, remains limited. Effort had been made to restrict international data collection for Australian species to regions climatically and edaphically comparable with the Philippines. However, only four farm forestry data points for *A. mangium* were found; hence the search was extended to include areas that cannot be regarded as directly comparable with the Philippines. West (1999) warned that this could introduce large errors into yield estimates; however, it is believed that a 'ballpark' figure is superior to no estimate.

Few smallholder forestry data were available, so height, diameter at breast height (dbh) and yield measurements were collected for a total of 163 representative or average trees in industrial plantations from the sources listed in Table 6.1. To predict yield for smallholder plantations, it was decided that the yield model would be fitted to the industrial data and an appropriate reduction factor would account for differences in management of farm plantations. Only two studies comparing growth and yield between plantations grown under farm and industrial management practices were identified. Adlard (1992) found a 41% difference in yield between industrial and farm plantations of *E. camaldulensis* in Karnataka, India, 10 years after establishment. On the advice of Thai forestry experts, Niskanen (1998a) applied a reduction factor of 25% to the mean annual increment in timber volume of industrial *E. camaldulensis* plantations in Thailand to approximate farm forestry yields.

6.4 DEVELOPING SMALLHOLDER YIELD MODELS FOR AUSTRALIAN SPECIES IN THE PHILIPPINES

Models were required to predict yield of fuelwood, rural house poles and sawlogs. The most valuable of these products are sawlogs, with a small-end diameter under bark (sedub) of at least 20cm. Their use in the Philippines is mainly for construction timber, but also includes plywood and veneer manufacture. Rural house construction poles are up to approximately 20cm diameter under bark (dub), and are less valuable than sawlogs. Electricity poles have not been considered, because of the longer time period required for trees to reach the desired size, which excludes this product as a viable farm forestry crop (Okit, 1999). Fuelwood includes prunings and logging residues. For this study, fuelwood is all timber less than 10cm dub.

Table 6.1 Sources of international data on height, diameter and yield for eucalypt species

Source	Data origin	Source	Data origin
Fenton *et al.* (1977)	World-wide	Wenlong and Jiayu (1989)	China
Sharma (1978)	India	Tsai (1993)[1]	S-E Asia
Poynton (1979)	South Africa	Eldridge *et al.* (1994)	World-wide
Hussain and Abbas (1983)	Pakistan	Janmahasatien *et al.* (1995)	Thailand
Jones (1983)[1]	Malaysia	Dongyun *et al.* (1996)	China
Gwaze (1989)	Zimbabwe	Luz *et al.* (1996a; 1996b)	Brazil
Milimo (1989)	Kenya	Pinyopusarek *et al.* (1996)	Lao P.D.R.
Pinyopusarek (1989)	Thailand	Turvey (1996)[2]	Indonesia

Notes: 1. Contains data for *A. mangium*. 2. Contains data for *A. mangium* and Australian eucalypts. Sources not footnoted contain data for one or more of the Australian eucalypts.

Few individual-tree data are available for the Australian species in the Philippines or international literature, which effectively precluded the development of growth models or *individual-tree* yield models. However, for the purposes of this study, *whole stand* yield models are appropriate. These can be further classified as *process*, *empirical* and *dynamic* models.

Process stand yield models, which attempt to simulate stand growth by describing the main physiological processes that convert carbon dioxide, nutrients and water into biomass through photosynthesis, were found to require data not available in the Philippines. Empirical stand yield models should be formulated to behave in a biologically realistic way to provide acceptable yield estimates (Vanclay, 1994). It was thought that the scarcity of data would hinder the development of a biologically consistent model, hence a dynamic stand yield model was chosen. Dynamic models incorporate some elements of growth theory into the model on the basis of past experience and not necessarily because these features are observed in the specific data being analysed. Pienaar and Turnbull (1973) asserted that without incorporating such theory, there is seldom reasonable justification for extrapolation beyond the conditions presented by the observed data. In this study the most popular dynamic model, the Chapman–Richards model, is used, where cumulative yield is represented by (Clutter *et al.*, 1983):

$$Y = \theta\{1 - \exp[-\phi(t - t_0)]\}^{[1/(1-\beta)]}$$

where Y = size of the organism or population (e.g. stand volume in m^3/ha);

$\theta = (\alpha / \gamma)^{[1/(1-\beta)]}$;

$\phi = \gamma(1 - \beta)$;
t = time (e.g. years) where $t = t_0$ when $Y = 0$; and
α, β, γ = constants ($\alpha > 0, \gamma > 0, 0 < \beta < 1$).

Test fittings of the Chapman–Richards model to data of each of the four eucalypt species revealed that the models for *E. tereticornis*, *E. pellita* and *E. urophylla* did not acceptably reflect observed yields or expert opinion for these species in the Philippines. Consequently, the scarce eucalypt yield data was pooled and a model for an 'average Australian eucalypt' developed.

Where height and dbh measurements were available for representative trees in stands from which foreign yield information was collected, individual-tree yield was re-estimated to the desired product specifications with taper functions. The taper function developed by Gordon *et al.* (1995) for *E. saligna* in New Zealand was adopted for eucalypts in the Philippines because it was developed from a large number of sample trees that closely resemble the likely ages and sizes of trees harvested in the Philippines. The lack of a suitable taper function for *A. mangium* necessitated a simplifying assumption; in this case that tree form is approximated by a cone with a base diameter equivalent to dbh and height equal to total tree height.

Table 6.2 presents the individual-tree yield and taper functions used in this study. Volumes of each product type for each species are estimated by setting dub in the taper function equal to the minimum small-end diameter for each product type and using the Microsoft Excel Goal facility to find the height that satisfies the equation. The log length and yield of each product type in the individual-tree are then estimated. When estimating yield per hectare, individual-tree volumes are assumed to be representative of the stand and are multiplied by the number of stems per hectare (sph). Where a measure of stand density was not reported, an initial planting density of 1,100sph (3m x 3m) is assumed with no scheduled thinning operations.

The Chapman–Richards model had been modified to accommodate non-linear panel data. This facilitated the use of a 'switch' to account for the expected differences in growth between industrial and small-scale forestry, while allowing data pooling. Prediction of yields for each management regime was not feasible for the average Australian eucalypt, because there were no data from smallholder management regimes. To account for the lower intensity management on smallholdings, a 33% reduction factor (the mean of the estimates by Adlard (1992) and Niskanen (1998a)) has been applied to local and foreign industrial yield data to generate the average Australian eucalypt model.

A Chapman–Richards model for each product type has been fitted to the stand yield data for *A. mangium* and the average Australian eucalypt by maximising the following log likelihood function (LF) with the Solver tool in Microsoft Excel:

$$LF(\theta, \phi, \beta, \sigma^2) = (-n/2)\ln(\sigma^2) - (n/2)\ln(2\pi) - 1/2\sigma^2 \sum (Y_{ij} - \theta\{1 - \exp[-\phi(t - t_0)]\}^{[1/(1-\beta)]})^2$$

where θ, ϕ, t, t_0, β are as previously defined;
n = number of yield observations;
σ^2 = variance of the yield data; and
Y_{ij}= timber volume per ha in stand i for product j.

Table 6.2 Individual-tree volume and taper functions for Australian eucalypts and A. mangium in the Philippines

Model	Austn. eucalypts (Gordon *et al.*, 1995)	*A. mangium*
Yield	$V = \pi / 40\,000 \times \int_0^{L1} [(\alpha_0 + \alpha_1 L/H + \alpha_2 (L/H)^8)(\beta_1 (L/H)^{(\gamma 1/H^{0.2})} + \beta_2 / (dbh\, H)^{0.3} (L/H)^{\gamma 2}]\, dL$	$V = SABH\,(H/3)$
Taper	$dub = \sqrt{\{(\alpha_0 + \alpha_1 L/H + \alpha_2 (L/H)^8) [\sqrt{(dbh^2 (\beta_1 (L/H)^{(\gamma 1/H^{0.2})} + \beta_2 / (dbh\, H)^{0.3} (L/H)^{\gamma 2})}]^2\}}$	$dub = dbh - ((dbh/H)(H - L))$
where:	$\alpha_0 = 0.8161$, $\alpha_1 = 0.09528$, $\alpha_2 = -0.2312$ $\beta_1 = [(1 - \beta_2 / (dbh\, H)^{0.3})(1 - 1.4/H)^{\gamma 2}] / [(1 - 1.4/H)^{(\gamma 1/H^{0.2})}]$ $\beta_2 = 4.298$, $\gamma_1 = 2.610$, $\gamma_2 = 30.72$ V = volume under bark (m^3) L_1 = total tree height when estimating total tree volume or distance below the growing tip if estimating less than total volume (m) H = total height of the tree (m) dub = specified diameter under bark, 10, 20 or other (cm) dbh = diameter at breast height measured at 1.4m above ground level (cm)	SABH = cross sectional area of tree bole at breast height (m^2) H = total height of tree (m). dub = specified diameter under bark, 10, 20 or other (cm) dbh = diameter at breast height (cm) L = distance from growing tip of tree to the specified diam. over bark (m)

The α, β, and γ parameters for the *A. mangium* model were allowed to vary between industrial and smallholder management regimes, i.e. $\alpha = \{\alpha_{m1}, \alpha_{m2}\}$, $\beta = \{\beta_{m1}, \beta_{m2}\}$and $\gamma = \{\gamma_{m1}, \gamma_{m2}\}$, where m_1 and m_2 refer to farm and industrial forestry, respectively. Venn and Harrison (2001) and Venn *et al.* (2001b) detail the modelling methods.

6.5 ADAPTING A FINANCIAL MODEL FOR FARM PLANTATIONS IN THE PHILIPPINES

A modified version of the Australian Cabinet Timbers Financial Model (ACTFM) (Herbohn *et al.*, 1998) has been integrated with the stand yield models to estimate the financial profitability of Australian species in small-scale plantations in the Philippines. Financial performance has been assessed in terms of land expectation values (LEVs) – useful for comparing species with short and differing rotations – and sensitivity analyses conducted on a range of input variables. Industrial yield models for the traditionally grown species *G. arborea* (Lingan, 1980), *P. falcataria* (Uriarte and Pinol 1996), *S. macrophylla* (Revilla *et al.*, 1976) and *E. deglupta* (Borough *et al.*, 1978) have also been integrated with the modified ACTFM to facilitate a comparison with the Australian species. A middle-range site index has been assumed for the traditional species.

During the 1970s and 1980s in the Philippines, real discount rates of up to 24% were employed for evaluating forestry investments (Appleton, 1980). More recently Uriarte and Pinol (1996) used 18% and DENR (1998a) applied rates of between 12% and 18%. In his study of Filipino and Thai forestry investments, Niskanen (1998a) adopted the rate of return required by international financing institutions such as the Asian Development Bank (currently around 10%–12%). Rates of 12% to 18% have been used here.

Prices for sawlogs of traditionally grown species in the Philippines were collected from various sources (DENR, 1998a; DENR, 1998b; ERDS, 1998; Magcale-Macandog *et al.*, 1998; Cuevas and Heath, 1999). Official information on prices for timber of Australian species is scarce, and none of the small-scale *A. mangium* growers visited had sold trees harvested from their properties. Consequently, the assumption is made that Australian eucalypt timber is comparable in value to the native *E. deglupta* at P3,000/m^3 for sawlogs. The market price for *A. mangium* sawlogs is assumed to be equivalent to *G. arborea* at P2,600/m^3, and P6,000/m^3 is used for *S. macrophylla*. All *P. falcataria* timber greater than 10cm dub is assumed to be used for house poles because the timber is unsuitable for sawlogs.

Rural house pole prices were obtained from ERDS (1998); P100 to P150 per 'average pole' is typical for a wide range of species. Assuming an average pole large-end diameter (led) of 17cm and length of 4m, use of the Gordon *et al.* (1995) taper function for eucalypts to determine pole volume implies a pole price of P2,000/m^3. According to Magcale-Macandog *et al.* (1998), farmers growing *G. arborea* at Claveria, Mindanao, sell fuelwood to bakeries in town for P1,200/tonne fresh weight or approximately P600/m^3. This is more than five times the reported fuelwood price in ERDS (1998). A compromise rate of P300/m^3 has been adopted for all species.

Management and cost data for industrial forestry with traditional species are available (e.g. DENR, 1998a; ERDS, 1998); however, these are unlikely to be relevant to small-scale forestry. Small-scale stand management practices were obtained from Filipino farmers for *A. mangium*, although the timing of operations was difficult to determine. No information was available for Australian eucalypts. Following the recommendation of Heath (1999), it is assumed that the types (but not timing) of silvicultural operations performed for all species grown in small-scale plantations can be approximated by BFI's industrial *A. mangium* and eucalypt plantation regimes (Cuevas and Heath, 1999). Table 6.3 reports this silvicultural information, including BFI's cost estimates, by species group for industrial plantations. This was used as the basis for small-scale plantation costs also presented in Table 6.3. On the advice of Aggangan (2000), a shadow wage for farm labour of P100/day (half the daily Philippine labourer wage) has been adopted. Assumptions for other smallholder silvicultural costs are as reported in Venn *et al.* (2001a). The species groups are *A. mangium* (m in Table 6.3), *G. arborea and P. falcataria* (gf), the average Australian eucalypt and *E. deglupta* (e), and *S. macrophylla* (my).

Initial stocking rates of smallholder plantings in the Philippines were observed at 1,100sph and 10,000sph. An average stand density of 2,500sph (2m x 2m) has been assumed. The timing of silvicultural operations for *G. arborea* and *P. falcataria* were taken from DENR (1998a). The timing of thinning and harvesting of small-scale *S. macrophylla* plantings are approximated by a regime reported by ERDS (1998), with timing of most other silviculture assumed to be the same as for eucalypts. It is assumed for small-scale plantings that: the site is prepared by brushing; natural fertilisers (e.g. manure) are used in place of chemical fertilisers; weed control is by hand; no thinning is performed; trees are pruned once only; and only half of the final crop is sold off-farm. The market price is applied to value timber used on-farm. Common practice is for *G. arborea* and *P. falcataria* plantations to be felled at age eight years (Magcale-Macandog *et al.*, 1998), *A. mangium* at nine years and eucalypts at 10 years (Niskanen, 1998a; White, 1988). A rotation of 20 years is assumed for *S. macrophylla*.

6.6 YIELD AND REVENUE PREDICTION

Table 6.4 reports predicted farm yields of the Australian and traditional species. Yields of Australian species are unexpectedly lower than those for traditional species, except in comparison with *S. macrophylla*. This is in contrast to industrial plantings where predicted yields of Australian and traditional species are similar (Venn and Harrison, 2001). It is to be noted that yields of traditional species assume a median site index, and that variations in yield of ±70% can arise due to assumed site quality.

Table 6.3 Assumed industrial and smallholder plantation silvicultural regimes and costs

Silvicultural operation	Industrial forestry		Farm forestry	
	Year m, gf, e, my	Cost (Pesos/ha)	Year m, gf, e, my	Cost (Pesos/ha)
Survey and mapping	0	500	na	na
Seedling production	0	1,600	0	1,600
Access and fire ponds	0	500	na	na
Land prep. – ind: burning, farm: brushing	0	250	0	1,000
Land preparation – hole digging	0	2,100	0	1,050
Land preparation – back-filling	0	2,200	na	na
Pest control	0	420	na	na
Land preparation – ground spray	0	2,200	na	na
Planting	0	3,300	0	1,650
Fertilisation 1	0	3,100	0	1,550
Fertilisation 2	1	3,000	na	na
Weeding – ind.: strip spraying, farm: hand.	1 & 2	1,200	1	1,875
Replacing dead trees	1	500	1	500
Form pruning	1	240	na	na
Second pruning	2, 2, 2, 3	360	2, 2, 2, 3	360
First thin	3	500	na	na
Supervision, management and administration overheads	a	1,800	a	200
Protection and maintenance of infrastructure	a	1,000	a	250
Contingency allowance	a	500	na	na
Second thin – logging and loading	7,na,10,17	28,800	na	na
Second thin – cartage	7,na,10,17	17,280	na	na
Second thin – roading and skid formation	7,na,10,17	5,760	na	na
Second thin – sales commission/management	7,na,10,17	5,184	na	na
Clearfall – logging and loading	13,8,15,25	92,160	9, 8, 10, 20	39,168
Clearfall – cartage	13,8,15,25	69,120	9, 8, 10, 20	29,376
Clearfall – roading and skid formation	13,8,15,25	11,520	9, 8, 10, 20	2,880
Clearfall – sales commission/management	13,8,15,25	17,280	9, 8, 10, 20	8,640

It is difficult to determine which site index (if any) from the *G. arborea* and *P. falcataria* models best approximates the site quality at ACIAR trial and smallholder sites in the Philippines. Possibly, a low site index may apply, in which case the predicted yields of Australian species will compare relatively more favourably with traditional species. Indeed, a critical advantage of growing Australian timber species that this study in yield modelling under scarce data fails to emphasise is that Australian species are comparatively well adapted to harsh sites. Farmers are likely to establish tree crops on their least productive land, which is unsuitable for establishment of many traditional species. Eucalypts, on the other hand, are '[o]ften . . . established on so-called 'wasteland' for rehabilitation of degraded sites where almost nothing else will grow' (Eldridge *et al.*, 1994, p. 3). Some of the international data collected for the Australian species may have provided poor approximations of the potential yield from plantations of these species in the Philippines. Additionally, the Australian species yield models were found to have an exponential form associated with the short rotations in the Philippines, which may have also affected the comparison.

Estimates of financial performance for small-scale plantations of Australian species are presented in Table 6.5, with traditional species included for comparison. At a discount rate of 18%, *P. falcataria* is predicted to have the highest LEV, followed by *E. deglupta* and *G. arborea*. *Acacia mangium*, Australian eucalypts and *S. macrophylla* all have predicted IRRs of about 18%, well below the other species in this study. However, it is possible that the assumed prices of Australian species sawlogs are conservative. This assertion relates to the comparatively low density and low to moderate natural durability of *E. deglupta*, *G. arborea* and *P. falcataria* timber (Williamson, 1993). *Acacia mangium* timber is denser than the latter two species and naturally more durable than all three species (Williamson, 1993). The timbers of the Australian eucalypts in this study are denser and more durable than the traditional species and *A. mangium* (Boland *et al.*, 1992). For uses in which strength and durability are prerequisite, including heavy engineering construction, railway sleepers, poles and flooring, especially under exposed conditions or areas subject to termites, Australian timber will out-perform timber from traditional plantation species. This could lead to premiums being paid for the timber of Australian species in the future. Sensitivity analyses have been performed with respect to a number of parameter values. Details are provided by Venn *et al.* (2001a). The sensitivity of LEV for each species was found to be most sensitive to changes (in descending order) to discount rate, site quality, rural house pole price, sawlog price, and clearfall logging and loading expense. These are all individually large cost and revenue items, or, in the case of the discount rate, affect all costs and revenues. A large range of LEVs due to variation in site quality, particularly for *P. falcataria* and *G. arborea*, reinforces the need for caution when applying the financial results.

Table 6.4 Predicted yields of Australian and traditional species grown under small-scale management regimes in the Philippines.

Item	*A. mangium*	Average Aust. euc.	*G. arborea*	*P. falcataria*	*S. macrophylla*	*E. deglupta*
Clearfall age (years)	9	10	8	8	20	10
Total yield (m^3/ha)	76	75	107	146	133	199
MAI (m^3/ha/year)	8	8	13	18	7	20
Sawlog yield (m^3/ha)	9	18	33	0	39	42
House pole yield (m^3/ha)	51	44	64	136	72	126
Fuelwood yield (m^3/ha)	16	13	10	10	22	31

Table 6.5 Predicted NPV, LEV, and IRR of smallholder plantings of Australian and traditional species in the Philippines

Item	*A. mangium*	Average Aust. eucalypt	*G. arborea*	*P. falcataria*	*S. macrophylla*	*E. deglupta*
NPV (pesos/ha) – 12%	5,800	8,000	42,700	65,900	18,400	70,000
– 15%	2,300	3,700	32,600	51,400	6,100	51,300
– 18%	-300	400	24,600	39,900	-900	37,200
LEV (pesos/ha) – 12%	9,000	11,800	71,700	110,600	20,500	103,300
– 15%	3,200	4,900	48,400	76,300	6,500	68,100
– 18%	-400	500	33,500	54,400	-900	46,000
IRR (%)	17.6	18.5	38.0	44.6	17.5	37.8

6.7 CONCLUDING COMMENTS

The Chapman–Richards yield models developed for smallholder plantings of Australian species predict a MAI of 8m^3/ha/yr for the eucalypts and *A. mangium*, which is considerably lower than estimates for *P. falcataria* and *E. deglupta* plantations. This is contrary to observations at ACIAR trials in the Philippines, where plantings of Australian species have been observed to grow more rapidly than stands of *E. deglupta*. The stand yield models developed for the Australian species appear to underestimate yield for short rotations typical of smallholder plantings. It is likely that the use of international data biased the yield models for Australian species, although the dearth of local yield data provided few alternatives in this study.

The financial analysis indicates that returns from small-scale plantings of Australian species are likely to be substantially lower than for plantings of traditional species. However, this analysis is constrained by a lack of yield and cost information for smallholder plantings and uncertainty surrounding market prices for timber of Australian species.

Sparse data for plantations of tropical species in developing countries is the rule rather than the exception. Preliminary yield and financial models, such as those presented in this chapter, can provide much useful information, despite data limitations. Preliminary models can also highlight important directions for future research. There is a need for further investigation into the growth performance and market acceptance of Australian species in the Philippines before passing recommendations regarding their suitability for planting or conducting extension programmes to support their use. Indications from this study are that some traditional species are likely to grow faster and provide higher returns in smallholder plantings on favourable sites, although Australian species are likely to have greater site adaptability and more desirable timber properties. It is anticipated that improved modelling, facilitated by future availability of growth and market data, will highlight the suitability of Australian species for small-scale forestry in the Philippines.

REFERENCES

Adlard, P.G. (1992), ‘Observed growth patterns of eucalypts in Karnataka’, in I.R. Calder, R.L. Hall and P.G. Adlard (eds), *Growth and Water Use of Forest Plantations*, New York: Wiley and Sons, pp. 63–73.

Aggangan, R. (2000), personal communication, Philippines Council for Agriculture, Forestry and Natural Resources Research and Development (PCARRD), Los Baños, Visayas.

Appleton, N.S. (1980), Yield Prediction Models and Economic Rotations of Unthinned *Albizia falcataria* (Linn.) FOSB. Plantations with Different Planting

Spacings, unpublished MSc Forest Resources and Management thesis, Los Baños: University of the Philippines.

Arnon, I. (1989), *Agricultural Research and Technological Transfer*, New York: Elsevier Applied Science.

Boland, D.J., Brooker, M.I.H., Chippendale, G.M., Hall, N., Hyland, B.P.M., Johnston, R.D., Kleinig, D.A. and Turner, J.D. (1992), *Forest Trees of Australia*, Melbourne: CSIRO Publishing.

Borough, C.J., Incoll, W.D., May, J.R. and Bird, T. (1978), 'Yield Statistics', in W.E. Hillis and A.G. Brown (eds), *Eucalypts for Wood Production*, Adelaide: CSIRO Publishing, pp. 201–25.

Clutter, J.L., Fortson, J.C., Pienaar, L.V., Brister, G.H. and Bailey, R.L. (1983), *Timber Management: A Quantitative Approach*, New York: Wiley and Sons.

Cuevas, E. and Heath, H. (1999), 'Direct and indirect cost of plantation timber/thinnings and profitability indicators at BFI', in S. Brown (ed.), *Kagubatan - The ACIAR Project 96/110 Newsletter March 1999*, Brisbane: The University of Queensland.

DENR (Department of Environment and Natural Resources) (1998a), *Development and Management of Forest Plantations: A Guidebook*, Laguna: Ecosystems Research and Development Bureau.

DENR (Department of Environment and Natural Resources) (1998b), *1997 Philippine Forestry Statistics*, Manila: DENR.

Dongyun, X., Guixiang, W. and Pegg, R.E. (1996), 'Value of selection in *Eucalyptus tereticornis* at Dongmen, People's Republic of China', in M.J. Dieters, A.C. Matheson, D.G. Nikles, C.E. Harwood and S.M. Walker (eds), *Tree Improvement for Sustainable Tropical Forestry. Volume 2*, Proceedings of the QFRI-IUFRO Conference, Caloundra, Queensland, Australia, 27 October – 1 November 1996, Gympie: Queensland Forestry Research Institute (QFRI), pp. 355–63.

Eldridge, K., Davidson, J., Harwood, C. and van Wyk, G. (1994), *Eucalypt Domestication and Breeding*, Oxford: Clarendon Press.

ERDS (Ecosystems Research Development Service) (1998), *Growth and Yield and Economic Analysis of Selected Forest Tree Species*, Techno-Transfer Series Bulletin No. 28, Davao City: DENR.

Fenton, R., Roper, R.E. and Watt, G.R. (1977), *Lowland Tropical Hardwoods: an Annotated Bibliography of Selected Species with Plantation Potential,* Wellington: External Aid Division of the Ministry of Foreign Affairs.

FWPRDC (Forest & Wood Products Research & Development Corporation) (1998), *Plantation Eucalypt Research, Making "The 2020 Vision" a Reality*, Canberra: FWPRDC.

Gordon, A.D., Lundren, C. and Hay, E. (1995), 'Development of a composite taper equation to predict over- and under-bark diameter and volume of *Eucalyptus saligna* in New Zealand', *New Zealand Journal of Forestry Science*, **25**(3), pp. 318–27.

Gwaze, D.P. (1989), 'Growth and survival of Australian tree species in field trials in Zimbabwe', in D.J. Boland (ed.), *Trees for the Tropics: Growing Australian Multipurpose Trees and Shrubs in Developing Countries*, Canberra: ACIAR, pp. 129–37.

Heath, H. (1999), personal communication, New Zealand Forestry Advisor at Bukidnon Forests Incorporated, Malaybalay, Mindanao.

Herbohn, J.L., Harrison, S.R., Emtage, N. and Smorfitt, D.B. (1998), 'The Australian Cabinet Timbers Financial Model for mixed species cabinet timber plantations in North Queensland', unpublished manual, Townsville: James Cook University.

Hussain, R.J. and Abbas, S.H. (1983), 'Provisional yield table of *Eucalyptus camaldulensis*', *The Pakistan Journal of Forestry*, **33**(2), pp.55–64.

Janmahasatien, S., Viriyabunch, C. and Snowdon, P. (1995), 'Soil sampling and growth prediction in Thailand', in T.H. Booth (ed.), *Matching Trees and Sites*, ACIAR Research Project No. 63, pp. 101–106.

Jones, N. (1983), 'Fast growing leguminous trees in Sabah', in Nitrogen Tree Fixing Association and IRDC, *Leucaena Research in the Asian-Pacific Region*, Proceedings of a workshop held in Singapore, 23–26 November 1982, Ottawa: IRDC, pp. 149–54.

Lingan, L.B. Jr. (1980), 'Growth and Yield Prediction for Yemane Plantations in the Philippines', unpublished MSc Forest Resources and Management thesis, Los Baños: University of the Philippines.

Luz, H.de F., Bierwagen, R., Brisolla, M. and Capitani, L.R. (1996a), 'Improvement of *Eucalyptus pellita* provenances and *E. resinifera* hybrids for board and timber production', in M.J. Dieters, A.C. Matheson, D.G. Nikles, C.E. Harwood and S.M. Walker (eds), *Tree Improvement for Sustainable Tropical Forestry. Volume 1*, Proceedings of the QFRI-IUFRO Conference, Caloundra, Queensland, Australia, 27 October – 1 November 1996, QFRI, Gympie, pp. 94–95.

Luz, H.de F., Bierwagen, R., Ferreira, M. and Capitani, L.R. (1996b), 'Population studies and selection strategies of *Eucalyptus urophylla*', in M.J. Dieters, A.C. Matheson, D.G. Nikles, C.E. Harwood and S.M. Walker (eds), *Tree Improvement for Sustainable Tropical Forestry. Volume 1*, Proceedings of the QFRI-IUFRO Conference, Caloundra, Queensland, Australia, 27 October – 1 November 1996, QFRI, Gympie, pp. 204–05.

Magcale-Macandog, D.B., Menz, K., Rocamora, P.M. and Predo, C.D. (1998), 'Gmelina timber production and marketing in Claveria, Philippines', in K. Menz, D.B. Magcale-Macandog, I.W. Rusastra, (eds), *Improving Smallholder Farming Systems in Imperata Areas of Southeast Asia: Alternatives to Shifting Cultivation*, ACIAR Monograph No. 52, Canberra: ACIAR, pp. 77–91.

Milimo, P.B. (1989), 'Growth and survival of Australian tree species in field trials in Kenya', in D.J. Boland (ed.), *Trees for the Tropics: Growing Australian Multipurpose Trees and Shrubs in Developing Countries*, Canberra: ACIAR, pp. 103–107.

Niskanen, A. (1998a), 'Financial and economic profitability of reforestation in Thailand', *Forest Ecology and Management*, **104**(1–3), pp. 57–68.

Niskanen, A. (1998b), *Cost-Benefit Analysis of Reforestation in Thailand and the Philippines*, unpublished PhD thesis, Joensuu: University of Joensuu, Faculty of Forestry.

Niskanen, A. and Saastamoinen, O. (1996), *Tree Plantations in the Philippines and Thailand: Economic, Social and Environmental Evaluation*, Research for Action 30, Helsinki: United Nations University, World Institute for Development Economics Research.

Okit, E.R. (1999), personal communication, Manager Forest Operations, Bukidnon Forests Incorporated, Malabalay, Mindanao.

Pienaar, L.V. and Turnbull, K.J. (1973), 'The Chapman-Richards generalisation of Von Bertanlaffy's growth model for basal area growth and yield in even-aged stands', *Forest Science*, **19**(1), pp. 2–22.

Pinyopusarerk, K. (1989), 'Growth and survival of Australian tree species in field trials in Thailand', in D.J. Boland (ed.), *Trees for the Tropics: Growing Australian Multipurpose Trees and Shrubs in Developing Countries*, Canberra: ACIAR, pp. 109–27.

Pinyopusarerk, K., Luangviriyasaeng, V. and Rattanasavanh, D. (1996), 'Two-year performance of Acacia and Eucalyptus species in a provinance trial in Lao P.D.R.', *Journal of Tropical Forest Science* **8**(3), pp. 412–23.

Poynton, R.J. (1979), *Tree Planting in Southern Africa. Volume 2. The Eucalypts*, Cape Town: Department of Forestry.

Revilla, A.V. Jr., Bonita, M.L. and Dimapilis, L.L. (1976), 'A Yield Prediction model for *Swietenia macrophylla* (King) Plantations', *The Pterocarpus*, **2**(2), pp. 172–79.

Sharma, R.P. (1978), 'Yield tables for Eucalyptus hybrid (plantation) for various levels of stocking', *The Indian Forester*, **104**(6), pp. 387–97.

Sy, M.U. (1998), 'Rehabilitation of natural logged-over forests: the Philippines scenario', *Canopy International*, November–December, pp. 9–11.

Tsai, L.M. (1993), 'Growth and Yield', in K. Awang and D. Taylor, *Acacia mangium: Growing and Utilisation*, Bangkok: Winrock International and FAO, pp. 149–61.

Turvey, N.D. (1996), 'Growth at age 30 months of *Acacia* and *Eucalyptus* species planted in *Imperata* grasslands in Kalimantan Seletan, Indonesia', *Forest Ecology and Management*, **82**(1–3), pp. 185–95.

Uriarte, M.T. and Pinol, A.A. (1996), 'Tree volume, yield prediction and economic rotation of *Albizia falcataria* in Mindanao, Philippines', *Journal of Tropical Forest Science*, **8**(3), pp. 289–99.

Vanclay, J.K. (1994), *Modelling Forest Growth and Yield: Applications to Mixed Tropical Forests*, Wallingford: CAB International.

Vendiola, E.E. (1996), Site-growth relationship of Acacia mangium Willd. The Case of Provident Tree Farms Incorporated (PTFI) Plantation in San Teodoro, Oriental Mindoro, unpublished PhD thesis, Los Baños: University of the Philippines.

Venn, T.J. and Harrison, S.R. (2001), 'Stand yield models for Australian eucalypts and acacias in the Philippines', in S.R. Harrison and J.L. Herbohn (eds), *Socio-Economic Evaluation of the Potential for Australian Tree Species in the Philippines*, ACIAR Monograph No. 75, Canberra: ACIAR, pp. 79-92.

Venn, T.J., Harrison, S.R. and Herbohn, J.L. (2001a), 'Relative financial performance of Australian and traditional timber species in the Philippines', in S.R. Harrison and J.L. Herbohn (eds), *Socio-Economic Evaluation of the Potential for Australian Tree Species in the Philippines*, Canberra: ACIAR, pp. 93-110.

Venn, T.J., Beard, R.M. and Harrison, S.R. (2001b), 'Modelling stand yield of non-traditional timber species under sparse data', in S.R. Harrison and J.L. Herbohn (eds), *Socio-Economic Evaluation of the Potential for Australian Tree Species in the Philippines*, ACIAR Monograph No. 75, Canberra: ACIAR, pp. 55-77.

Wenlong, Z. and Jiayu, B. (1989), 'Tropical eucalypt trials on Hainan Island, People's Republic of China', in D.J. Boland (ed.), *Trees for the Tropics: Growing Australian Multipurpose Trees and Shrubs in Developing Countries*, Canberra: ACIAR, pp. 79–87.

West, P.W. (1999), personal communication, Sci-West Consulting, Meerschaum Vale, New South Wales.

Williamson, M. (1993), *Forest Management in Industrial Forest Plantations in the Philippines*, Canberra: FORTECH.

White, K.J. (1988), '*Eucalyptus* on small farms', in D. Withington, K.G. MacDicken, C.B. Sastry and N.R. Adams (eds), *Multipurpose Trees for Small-Farm Use*, Proceedings of an International Workshop November 2–5 1987, Pattaya, Thailand, Arlington: Winrock International Institute for Agricultural Development and IDRC, pp. 86–96.

7. The Role of Portable Sawmills and Chainsaw Milling in Tropical Small-scale Forestry

D.B. Smorfitt, J.L. Herbohn and S.R. Harrison

Various forms of portable or 'walkabout' sawmills have become popular for timber extraction and in-forest milling in tropical forestry in recent years. A number of advantages over larger fixed-site mills have been recognised, such as reduced transport costs and utilisation of smaller logs and branches. Portable bandsaws, with a thin blade kerf generally give rise to a greater timber recovery than traditional circular saws. Chainsaws have been criticised for greater timber wastage, and in some cases their use for timber extraction has been declared illegal. The choice of appropriate technology involves a tradeoff with respect to various factors including capital outlays, maintenance capability and costs, labour efficiency and timber recovery rate. If these items are factored into the overall average cost curve, it becomes apparent that both portable sawmills and chainsaw milling do have an important role in tropical small-scale forestry. The latter are particularly useful for flitch cutting as a first stage of milling.

7.1 INTRODUCTION

The terms 'portable sawmills' and 'mobile sawmills' are frequently used interchangeably. Sometimes the distinction is made that portable sawmills can be dismantled and physically carried to inaccessible areas, while mobile sawmills are those transported between sites by vehicle, animal, or human means, so that logs can be milled at source. The essential difference is one of portability and thus greater access to the forest. Larger mobile sawmills may be found in smaller family-owned traditional fixed-site mills, frequently used as the initial saw to break the logs into quarters before they are moved to the resaw benches.

Early use of portable sawmills is reported in Brazil in 1968 (Hasek and Ponce, 1973). Vast quantities of timber were being destroyed for settlement, road construction and other development activities, with no commercial sawmills existing within an economic recovery range. It is in this context that portable sawmills first received attention. From its beginnings in 1962, the Mobile Dimension portable sawmill has gone on to achieve sales exceeding 4,000 units, and has been sold in 50 states in America and in 60 other countries (Fisher, 1986). Initially, portable sawmills predominantly operated in developing countries including Brazil, Papua New Guinea, Solomon Islands and Vanuatu. According to Bun (1993) more than 600 'Wokabout Somils' have been sold in Papua New Guinea alone. Sale numbers for portable sawmills appear to be sufficient for them to play a significant role in the timber industry in developing countries.

Increased usage of portable sawmills may be attributed to a number of factors, including: greater awareness of the availability of these mills; improved marketing and promotion by manufacturers; greater promotion and public awareness of the potential environmental 'friendliness' of these sawmills; claims of higher recovery rates achievable; and attempts by landholders to find ways of supplementing incomes, and reducing risks faced, by diversifying into another, yet complementary field.

The chapter examines the advantages of portable sawmills as opposed to traditional fixed-site mills. The advantages of portable sawmills and the various types and their pros and cons are outlined. Cost information on ownership and operation of these mills is provided.

7.2 ADVANTAGES OF PORTABLE SAWMILLS

The following potential advantages of portable sawmills over fixed-site mills have been identified (Kowero *et al.*, 1985; Department of Primary Industries – Forestry, 1998):

- They are relatively inexpensive to acquire;
- They can increase the amount of sawn timber recovered from a given log relative to larger fixed location mills;
- Production capacity can commence without the delays associated with planning and construction of the sawmill;
- Logs can be sawn where they are felled thereby reducing transport through reduced volume of waste and more organic matter being left in the forest;
- Portable sawmills can handle small diameter logs, such as those from thinnings and larger branches;

- They are more flexible in their operation, for example cutting small and odd sizes with minimal re-setting of equipment; and
- Portable sawmills are not as demanding in terms of technical skills as fixed-site sawmills.

Other advantages reported by Stewart and Hanson (1998) include landholders taking advantage of niche markets that industrial scale processors cannot meet, creating employment in rural areas, farmers milling in the quieter off-seasons, and reduced damage to the forest. Another reported advantage is reduced roading cost and environmental impact.

The relative importance of these advantages will vary quite considerably depending on specific circumstances. In-forest milling imposes extra costs associated with either provision of on-site accommodation or daily travelling to the site. Also, while it may be physically possible to mill small diameter logs, milling costs will increase and the timber may have low value. Generally, portable sawmillers, like their fixed-site counterparts, would prefer large diameter logs if given the choice.

7.3 CHARACTERISTICS OF PORTABLE SAWMILLS

There are several types of portable sawmills, manufactured in various countries including Canada, United States of America, Australia, New Zealand and Finland. The nature, operation and size of these mills can differ considerably.

Chainsaw Mills

Chainsaw mills are the least sophisticated of the portable sawmills. These mills consist of a chainsaw attached to a lightweight metal frame that is positioned over the log. Attachments for the frame are available which allow the chainsaw, with the addition of a longer bar, to mills slabs. These mills are generally inexpensive and easy to operate. Due to the wide kerf, these mills have low timber recovery rates. Also they require considerable manual handling of logs, have low throughput and produce rough sawn surfaces. They are generally only suited to low-value low-volume milling or as a means to break down logs for further sawing.

Chainsaw mills do however have potential for use in situations where portability, low cost of acquisition, and low environmental impact are priorities. Farmers generally own a chainsaw and thus, with a little additional cost to purchase the lightweight frame, are able to mill small volumes of wood for farm use. This allows the utilisation of logs that might otherwise be discarded, for farm applications such as cattle yards and sheds.

For example in New Zealand, the Forest Research Institute (1985) reported a 'new' chainsaw mill that was used to mill nine large 400–500 year old rimu growing in a dense stand of 100–150 year old poles in the State Forest. Normally these larger trees would not be harvested due to the danger of damage to younger trees. The trees were felled into canopy gaps and the nature of the milling equipment only required a narrow workspace around each log. This study reports a kerf of approximately 9mm and most boards oversized by about 5mm to allow for dressing. Conversion from sawlog to clearwood pieces was 52% of which 27% was heart dressing grade, 64% dressing grade, and 9% building grade. This high recovery rate was obtained from milling exceptional logs, and could not be expected with small diameter logs and smaller dimensioned boards.

Chainsaw mills play an important role in developing countries where they are used to mill logs into large flitches so that transport of waste material is reduced to a minimum. Flitches are then transported to larger mills for sawing into boards. Another application is to mill high value timbers in the forest and carry the boards out to a central location. This will only be possible in countries where labour is inexpensive. Due to their low environmental impact, another use is for selective logging of native forests where other values such as scientific and wildlife value outweigh timber values. Extraction and removal of only the high value timber involves little loss of nutrients and additional seedlings can be planted to promote regeneration (Forest Research Institute, 1985).

Bandsaws and Circular Saws

A wide variety of bandsaw and circular saw mills are available. The relative suitability of circular saws and bandsaws for milling high quality boards for cabinet making and the recovery rates achievable is an ongoing debate between proponents of each mill type. Folkema (1992) reported that circular sawmills have long been used by Canadian farmers but sales have dropped since the introduction of mobile bandsaw mills in 1982.

Portable sawmills may be powered by electric, petrol or diesel motors. If the portable sawmill is to be moved from site to site, a diesel or petrol motor is required. Petrol motors typically are used on smaller sawmills, whereas diesel motors are used on the larger commercial portable sawmills.

Circular saw mills may use a single, double or multi-blade cutting configuration. A double blade configuration features a vertical and horizontal blade that operate at the same time, cutting two sides of planks with one pass over the log. With a single blade configuration, orientation may be switched between vertical and horizontal between passes.

Folkema (1992) has identified the following advantages of each mill type. Advantages of portable circular sawmills include:

- Less blade sharpening and blade maintenance is required;
- Output tends to be up to 50% higher for one or two man operations; and
- Some timber species with dense knots are easier to saw because circular saw blades are thicker and more rigid. (This could be applicable for example to eucalypt hardwoods.)

Advantages of portable bandsaw mills include:

- The kerf on a bandsaw blade is about one quarter that of a circular saw blade so timber yield is higher;
- They can usually manage larger diameter logs than circular saws;
- They are generally safer and quieter to operate;
- They require about a quarter to a third of the power required to operate a circular saw; and
- Where there are nails and other obstructions causing blade breakage, replacement cost and downtime are generally lower than with a circular saw.

Relative economic performance of portable bandsaw and circular saw mills depends critically on timber recovery rate and log throughput. Recovery rate is the amount of sawn timber output in cubic metres expressed as a percentage of the roundlog input in cubic metres. Bandsaw proponents argue that higher recovery rates are achievable for a number of reasons, and that bandsaws are more suitable for milling logs into a variety of dimensions, and are also highly appropriate for cutting a whole log into a number of slabs (bark to bark). Circular saw proponents point to the higher throughput achievable with circular saws, and refer to the wave effect of the wandering bandsaw blade resulting in further costs associated with dressing of sawn boards. Bandsaw proponents claim a well-cut board from a bandsaw should need little dressing, and that the wave effect is due to poor maintenance of the guides on the mill and a lack of experience of the operator.

7.4 ISSUES IN THE DEBATE BETWEEN PORTABLE AND FIXED-SITE SAWMILLS

A number of questions relating to portable sawmills deserve further consideration. These include: are portable sawmills complementary to, or competitive with, fixed-site sawmills? What role do portable sawmills have to play in the current timber industry? Is the role limited to hobbyists and farmers cutting for their own requirements? Is the role limited to emerging or diminishing markets where log resource availability is too low to warrant commercial sawmilling? Answers to these questions will be very much dependent upon the circumstances in which the sawmill is operating. For instance the answer may differ where labour is inexpensive to where high labour costs are encountered.

Recovery rate depends on ability, experience and skill of the miller, kerf of the blade, volume of timber cut, species of timber cut, dimensions of the sawn timber cut, and quality and dimensions of the log input. Recovery rates have a major impact on cost structures and revenue achievable by sawmillers. An increase in recovery rates from 35% to 40% for a firm paying $\$50/m^3$ stumpage for roundlog inputs, is equivalent to a decease in the sawn timber equivalent input price from $\$143/m^3$ to $\$125/m^3$

Recovery rates are a contentious issue. Some promoters of portable sawmills claim recovery rates of 60% to 70%, which is considerably higher than the approximate 35% (Bennett, 1990; Queensland Timber Board, 1996) achieved by fixed-site mills. Recovery rates achieved by fixed-site sawmills in the Wet Tropics rainforests of North Queensland in 1987 were on average 39.8% (Cameron McNamara Consultants, 1988).

Gan *et al.* (1985) quote recovery rates for portable sawmills ranging from 22% to 47% depending on cultivars, with a mean ratio for nine cultivars of only 32%. Large differences between recovery rates were found for different Heveawood (rubberwood) cultivars. It is notable that these trees were not originally planted for sawn timber.

Hasek and Ponce (1973) report 'The loss was negligible and the yield high, approximately 64%, resulting from the good quality and excellent condition of the log.' It is notable that a log of lower quality would not necessarily have produced a similar recovery rate. Mamum and Konabe (1994) report that a successful portable sawmill operated by a family group in Papua New Guinea consistently achieved a rate of recovery of more than 55% and a daily production capacity in excess of two cubic metres.

Fixed-site sawmillers in Australia on the other hand argue that high recovery rates are generally not feasible under the conditions in which they operate, where selective harvesting is not possible and 'everything' is harvested. Recovery rate is also reduced where a large number of timber species are harvested. This was the case in North Queensland while logging of rainforests was still being undertaken. For example, Poore (1988) indicates that there are approximately 800 tree species in the rainforest[1] of which about 600 reach sawlog size, and as many as 150 of these species

[1] Hyland and Whiffin (1993) indicate that there are 1,056 taxa present as trees in Australian tropical rainforests.

were harvested. When sawmillers purchased the main species in demand, such as Red Cedar, Queensland Maple and Silky Oak, they were also expected to take lesser-known species.

A major problem faced by portable sawmill operators is marketing their output. Stewart and Hanson (1998) have highlighted the problem that many portable sawmillers overlook marketing during the planning stage. They also note that the lack of information on marketing of sawn timber by small-scale processors. Trees can be grown and milled into boards, but it is necessary to find a market for the product. Timber grower co-operatives may be a useful tool in the marketing of timbers that originate from disparate sources. Other problems faced by portable sawmillers are the drying and storage of timber and the treatment of timber against insect attack. The importance of supplying dried timber to cabinet-makers who work on a 'just-in-time' inventory basis and thus carry minimal solid timber stocks has been identified by Herbohn *et al.* (1997)

7.5 CASE STUDIES IN APPLICATIONS OF PORTABLE SAWMILLS

Portable Sawmills in Australia

The extent of utilisation of portable sawmills is sometimes difficult to gauge. In Queensland sawmills have to be licensed with the Primary Industries Corporation in terms of the *Sawmill Licensing Act 1936*. There are currently 112 portable sawmills licensed (1996–97), with some licensed portable sawmills operating at fixed locations. In addition to licensed portable sawmills, there are many of the smaller millers that do not fall within the definition of a sawmill as per the *Sawmill Licensing Act 1936*. Anecdotal evidence suggests that there are approximately 300 unlicensed portable sawmills of which approximately 100 are accredited with the Australian Portable Sawmill Association Pty. Ltd. (Goodman, 1998). FORTECH (1994) notes the lack of formal sources of statistics on portable sawmilling including output, cost and timber quality.

Laidlaw (1997, p. 71) indicates that 'Forestry Tasmania does not permit portable sawmilling of any kind on State forest in Tasmania. This is because such an activity is a potential source of fire and portable sawmilling is very difficult to regulate.'

Training and development programmes that enhance skills and abilities are of particular importance for new entrants into sawmilling, where significant dangers associated with working with high speed cutting equipment such as chainsaws and portable mills are prevalent. A recent entrant to the portable sawmilling industry is the Australian Portable

Sawmill Association Pty. Ltd. This organisation provides training which, based on information supplied, is nationally recognised by the Vocational Education, Training and Employment Commission (VETEC) of Queensland and has Australian Portable Sawmill Association (APSA) endorsement. The course covers portable sawmill and chainsaw operation, wood properties and uses and visual stress grading, A Forest Industry Training and Education Consortium subsidy is available in some instances (FITEC, 1997).

Stewart and Hanson (1998) indicate that only two of the six Australian sawmill manufacturers covered in their report offered training as part of the purchase package or as an optional extra. The Victorian Timber Industry Training Centre in Creswick also offers occasional training in this area. Stewart and Hanson (1998) indicate that 16 of the 25 portable sawmillers interviewed had no previous sawmiling experience, and only six millers had undertaken any training. Thus it appears to be a matter of learning on the job, which may not only prove to be expensive financially due to poor quality output and higher levels of waste, but also opens the millers to potential accidents.

It has been suggested that portable sawmills have the capacity to replace fixed-site sawmills in North Queensland due to the fragmented and depleting resources. High throughput volume may result in lower unit milling costs per cubic metre of sawn timber. This may well be the case where the sawmill is milling standardised species, log size and log quality from plantation-grown softwoods, but need not be the case for milling of mixed species of varying quality and dimensions (Smorfitt, 2000)

Portable Sawmills in Developing Countries

In developing countries an alternative to large-scale commercial milling is based on low technology methods. Wyatt (1998) reported that many South Pacific countries have initiated programmes to investigate or promote portable sawmills with common objectives for these programmes including:

- Using portable sawmills to encourage rural development through on-site value adding, employment creation and providing timber for local requirements;
- Reducing forest harvesting levels by enabling local communities to meet income requirements through lower log volume, with high-value sawmilling replacing low-value, high-volume logging; and
- Promoting sustainable long-term forest management by giving local communities greater appreciation of the forest value and providing an effective means of management.

Following is a brief outline of the role of portable sawmills in a number of developing countries.

Papua New Guinea: Accessibility of advice, spare parts, servicing and training has proven to be a problem with portable sawmills in remote areas in developing countries such as Papua New Guinea. Limited access to training and spare parts has resulted in approximately 50% of the Wokabout Somils being non-operational (Padua, 1994). This situation does not appear to have improved in recent years with Hunt (1999) indicating that many of the portable sawmills in Papua New Guinea are not operational. Hunt *et al.* (1999) suggest that many of the groups in Papua New Guinea will be part-time producers operating intermittently depending on the demands of other economic and cultural pursuits and the need for cash.

Vanuatu: Wyatt (1998) indicates that his experiences in Vanuatu resulted in him changing his promotion of portable sawmills to one of promoting chainsaw milling as they were more appropriate to the needs of the people. The significantly lower costs enabled the people to remain part-time sawmillers milling timber on an occasional basis when timber or cash was required. Wyatt does acknowledge the wastage argument but suggests that chainsaws may be an efficient manner in which to meet the needs of people living near the forests and involving them in forest management.

Philippines: Personal observation of the authors reveals that portable sawmills are used extensively in very small-scale forestry in the Philippines, such as small woodlots and fenceline plantings. Coconut plantations also provide a source of occasional logs, although restrictions are now in place over the felling of coconut trees. In some cases, farmers deliver sawn gmelina, mahogany and cocolumber timber to lumber merchants, sawing being by chainsaw or bandsaw mills. Also, employees of small lumber merchants go to farms and fell and flitch trees by chainsaw. The price is paid to farmers on the basis of the sawn timber recovered. This shifts the cost of low recovery rates from the merchants to the landholder. Sawing to planks is then carried out on small bench circular saws at the lumber yard when workers have uncommitted time. Regulations restricting use of band and circular sawmilling equipment has lead to increased use of chainsaws.

Brazil: d'Oliveira *et al.* (1998) reports that in Brazil portable sawmills and chainsaw mills are used to mill timber. Extraction of timber is frequently initially for local housing and community structures and then for local timber markets and those in the closest cities. The extraction is often done on a seasonal basis, with dry weather harvesting being preferred. However, other activities like hunting, fishing and harvesting non-timber forest products along with subsistence agriculture also influence the timing of harvesting (d' Oliveira *et al.,* 1998).

Bolivia: Kress (1998) reports that in Bolivia, legislation regulating harvesting of forests specifically forbids the chainsaw milling of timber. The new Forestry Law of Bolivia reportedly stipulates that 'Under no circumstances shall the use of chainsaws be allowed for the squaring and sawing of timber for commercial purposes' (Article 75). In Bolivia the sawmilling sector is characterised by under-capitalisation, small size, utilisation of old equipment, intensive labour use and low productivity. Almost all sawmills require investment and few have appropriate cost management planning systems or employ rational criteria for determining investment (King *et al.,* 1996).

Peru: In the cloud forests of Peru, the steep terrain and low to medium soil quality have necessitated a 'reduced-impact logging plan' be designed (Rebaza, 1999). Milling begins in the forest with the quartering of timber to planks with chainsaw-based portable sawmills alongside the stump. The planks are hand-carried to the nearest road at which point they are placed on mules for the further transport to the Timber Processing, Training and Production centre. Rebaza (1999) suggests that the advantages of these modes of transport include minimal disturbance caused to the natural environment, their simplicity and adaptability to topographic conditions, and the lack of a need for major investment in road construction and logging equipment. At the Timber Processing Centre, a band re-saw mill has been installed to allow for further milling. This timber together with timber products made at the centre are further transported by mule and road to the Furniture Assembly and Sales Centre. Here the furniture is assembled and the finishing touches added prior to sale.

7.6 CONCLUDING COMMENTS

The role portable sawmills can play in the sawmilling industry differs according the specific circumstances, but a role they can definitely play. Throughout the industrialised and developing world, chainsaw, circular saw and bandsaw mills are being utlised to meet the demands of niche markets and local communities. Whilst they may not be able to compete directly with large fixed-site mills that mill standardised plantation timbers, they most definitely can complement these fixed-site mills. Their use is applicable in instances where log resource is scattered, there is a lack of road infrastructure, where conventional milling may not be advisable due to scientific or environmental values exceeding timber value.

REFERENCES

Bennett, J. W. (1990), *Commission of Inquiry into the Conservation, Management and use of Fraser Island and the Great Sandy Region*, Appendixes Volume 1, Brisbane: Goprint.

Bun, Y.A. (1993), 'Report on the nationwide wokabout somil survey', *Klinkii*, **5**(1), pp. 34–36.

Cameron McNamara Consultants (1988), *Socio-economic Study: Wet Tropical Rainforests North Queensland*, Commissioned by Queensland Government Premier's Department, Brisbane.

d'Oliveira, M.V.N., Braz, E.M., Burslem, D.F.R.P. and Swaine, M.D. (1998), 'Small-scale natural forest management: a new model for small farmers in the Brazilian Amazon', *Tropical Forest Update*, **8**(1), pp. 5–7.

Department of Primary Industries – Forestry (DPI Forestry) (1998), *An Overview of the Queensland Forest Industry*, Information Series QI 98015, Brisbane: DPI Forestry.

Fisher, R. (1986). 'If the trees won't come to the sawmill ....', *The Business Journal*, **6**(1), pp. 1–3.

Folkema, M.P. (1992). *Handbook on Portable Bandsaw-type Sawmills*, Forest Engineering Research Institute of Canada (FERIC) Handbook No. 10, Pointe Claire, Canada: FERIC.

Forest Industry Training and Education Consortium (FITEC) (1997), *Training Information, What are my Training Options?*, Forestry Industry Training and Education Consortium, information brochure, Canberra.

Forest Research Institute (1985), 'A new portable chainsaw mill', *What's New in Forest Research*, No.132, Rotorua, New Zealand: Forest Research Institute.

FORTECH (1994), *Commercial Potential for Plantations of High Value Rainforest Timbers in North Queensland – Assessment of Market Opportunities for Plantation Grown Wood Products*. Report prepared by Forestry Technical Services Pty. Ltd. for the North Queensland Afforestation Project Joint Board, Canberra: FORTECH.

Gan, L. T., Ho, C.Y. and Chew, O.K. (1985), 'Rubberwood: sawntimber production and recovery studies', in K.D. Singh, C.K. Ten, and H.L. Thong (eds), *Proceedings of the Second Rubberwood Seminar*, Kuala Lumpur, Nov 19–20, 1985, Kuala Lumpur: Forest Research Institute Malaysia, pp. 97–122.

Goodman, M. (1998), personal communication, Australian Portable Sawmill Association Pty. Ltd, Yeppoon.

Hasek, V.C. and Ponce, R.H., (1973), 'Dimension Sawmills', *Brazil Florestal*, **4**(14), pp.47–60, translation from Portuguese, Wellington: Department of Internal Affairs.

Herbohn, J. L., Smorfitt, D.B., Peterson, R. and Harrison, S.R. (1997) *Cabinet-Making Firms and the Demand for Rainforest Timbers: (I) Townsville Survey*. Research Report Series No. 7. Department of Accounting and Finance, James Cook University of North Queensland, Townsville: Cooperative Research Centre for Tropical Rainforest Ecology and Management.

Hunt, C. (1999a), personal communication, Senior Research Fellow, The National Research Institute, Port Moresby, Papua New Guinea.

Hunt, C., Chatterton, P., Bun, Y., Whimp, K. and Eddows, P. (2000), A Future For Our Forests: Strategies for community-based forestry and conservation in Papua New Guinea, WWF, Suva.

Hyland, B.P.M. and Whiffin, T. (1993), *Australian Tropical Trees - An Interactive Identification System*, Volume 2, Melbourne: CSIRO Publications.

Kowero, G.S., Kalage, S.E. and O'kting'ati, A. (1985), 'The performance of mobile sawmills in Northern Tanzania', *Commonwealth Forestry Review*, **64**(3), pp. 219–225.

King, K. *et al.* (1996), 'Future directions for Bolivia', *Tropical Forest Update*, **6**(4), pp. 14–16.

Kress, A. (1998), 'Promoting chainsaws', letter to the editor, *Tropical Forest Update*, **8**(3), p. 30.

Laidlaw, V.D. (1997), *The International Portable Sawmillers Guide*, 1st Edn., Cudgee, Victoria: V.D. Laidlaw.

Mamum, J. and Konabe, C. (1994), 'Experiences with wokabout somil (mobile sawmill) in Papua New Guinea – prospects and problems', *Papua New Guinea Forest Research Institute Newsletter*, **1**(6), pp. 3–6.

Padua, A.I. (1994). 'The economic impact of wokabout somil in the forest industry of Papua New Guinea', *Papua New Guinea Forest Research Institute Newsletter*, **1**(6), pp. 7–8.

Poore, D. (1988), *Natural Forest Management for Sustainable Timber Production*, Report prepared for the International Tropical Timber Organisation, PPR 11/88(F), London: IIED.

Queensland Timber Board (1996), *The Queensland Timber Industry - an overview*. Information note, Brisbane: Queensland Timber Board.

Rebaza, M.P. (1999), 'Demonstration community forest management in the natural cloud forests of Peru', *Tropical Forest Update*, **9**(2), pp. 12–13.

Smorfitt, D.B. (2000), A Commercial Evaluation of the Sawmilling Industry in North Queensland with Special Reference to the Current and Potential Role of Portable Sawmills, M.Com. Thesis, James Cook University, Townsville.

Stewart, M. and Hanson, H. (1998), *On-site Processing for Farm Forestry*, Rural Industries Research and Development Corporation (RIRDC), Publication No. 98/79, Canberra: RIRDC.

Wyatt, S. (1998), 'Chainsawmilling', Letter to the Editor, *Tropical Forest Update*, **8**(4), p. 30.

8. Choice of Timber Inputs by Small to Medium Sized Cabinet-making Firms in Queensland and Implications for the Marketing of Lesser-known Tropical Species

J.L. Herbohn, D.B. Smorfitt and S.R. Harrison[1]

This chapter discusses the marketing of tropical rainforest cabinet timbers, particularly in relation to small to medium sized cabinet-making firms. While this analysis is set in context of Queensland firms and Australian rainforest and eucalypt species, many of the principles can be applied to other tropical species. Timber availability, suitability, customer request and colour and grain are the most important factors in the decision of cabinet-makers to select a particular species. Price only becomes important when it cannot be passed on to the purchaser. Australian rainforest cabinet timbers are highly regarded by both cabinet-makers and the general public; however, their use is restricted by actual and perceived availability. Queensland maple, red cedar and northern silky oak are the rainforest species most highly recommended by cabinet-makers for planting to satisfy their future timber requirements.

8.1 INTRODUCTION

The use of North Queensland rainforests for timber production became a controversial environmental issue from the early 1960s, with calls for conservation of particular areas becoming ever more frequent in the 1970s (Carron, 1993; Adam, 1994). After heated public debate over the logging in

1 We wish to thank the Rainforest CRC and North Queensland Joint Board for providing funding for this project. The assistance of Cr Mike Berwick, Mr Jim Meharg, Mr Bill Sokolich, Mr Trevor Ritchie and Mr Gary Sexton when developing questionnaires for this study, and of Mr Nick Emtage in statistical analysis is also gratefully acknowledged.

the early 1980s, the Commonwealth Government unilaterally nominated rainforests on Crown land in North Queensland for World Heritage listing in December 1987, despite vehement opposition from the Queensland State Government and local government authorities. World Heritage listing resulted in complete cessation of logging activities on all Crown rainforest areas in North Queensland, and a dramatic decrease in the availability of rainforest cabinet timbers. Current supplies are sourced almost exclusively from freehold rainforest areas not included as part of the listing.

Moore (1992) has summarised rainforest timber removals from Crown and private land immediately prior to World Heritage listing. The three-year average harvest for rainforest cabinet timbers was 66,000m^3, with a further harvest of 50,000m^3 of lower grade rainforest structural timber (used in the building industry). In 1987, the year prior to listing, just over 63,000m^3 of cabinet woods were harvested, 52,000m^3 from Crown land. After listing, the Crown rainforest harvest dropped to virtually nil and the private cut increased to about 30,000m^3 by 1992 (Moore, 1992). The cut from private land has continued due to establishment of new sugarcane areas.

Recent unpublished market data indicate that premium quality species (seasoned and dressed) such as Queensland maple and northern silky oak characteristically retail for approximately A$2,000 /m^3, similar to that prior to World Heritage listing.

Retail prices of rainforest cabinet timbers (RFCTs) are also often similar to, or lower than, those being obtained for equivalent imported tropical species. This is in agreement with the observations of Moore (1992) that price decreases of Australian RFCTs occurred as a consequence of reduced availability and irregularity of supply causing a substitution with other species or disappearance of the end-users of the timbers. Such anecdotal evidence suggests that availability and regularity of supply may be important determinates of use and price.

This chapter reports the results of surveys of cabinet-makers and the general public in Cairns, Townsville and Brisbane. These surveys were designed to provide insights into the important factors affecting the demand for rainforest cabinet timbers and the choice of species by cabinet-makers and their customers.

8.2 SOURCES OF DATA

Three separate but interrelated surveys have been conducted that provide information about the marketing of rainforest cabinet timbers in Queensland. The first survey was of managers of cabinet-making firms in Cairns, Townsville and Brisbane. The questionnaire comprised three sections: general information on mangers and the characteristics of their firms; questions relating specifically to the use of rainforest cabinet timbers and a range of other timber inputs in cabinet-making operations; and multiple-part questions asking managers their opinions on why people buy, or do not buy, products made from RFCTs. During interviews, managers also had the opportunity to list further factors that had not been included in each series of statements. Further details of the sample selection process and the research methods applied can be found in Herbohn *et al.* (1997), Smorfitt *et al.* (1997) and Peterson *et al.* (1997).

The second survey was of the employees (as distinct from managers) of the same cabinet-making firms, and solicited views on what species should be planted to satisfy future timber requirements of the firms. Questions were included on awareness, usage, and suitability for three purposes, viz. furniture, kitchen benchtops and kitchen cabinet doors. The third survey investigated attitudes of the general public to products made from rainforest cabinet timbers and the reasons why they have purchased or not purchased products made from rainforest cabinet timbers. Data were also collected on when respondents last purchased products made from these timbers. Personal interviews and telephone surveys using a prepared questionnaire were conducted of the general public at several locations in Cairns and Townsville during 1997. Two methods were used for obtaining sample members. The first involved a convenience sampling approach of intercepting members of the general public at shopping malls. The second comprised a telephone survey of respondents randomly selected from the Townsville Telstra White Pages. Adult respondents (over 18 years) were sought in both surveys. Further details of the research methods and sample selection process of these two surveys can be found in Smorfitt *et al.* (in process, b).

8.3 PROFILES OF QUEENSLAND CABINET-MAKING FIRMS

As indicated in Table 8.1, furniture construction comprises a large proportion of the product mix of firms with low turnover (<$100,000 pa.) in each city. In contrast, construction of kitchens comprises a much higher proportion of the product mix of medium ($100,000 to $300,000 p.a.) and large firms (> $300,000 pa). Materials and labour each comprise about 40% of the cost of products manufactured by cabinet-making firms, with unspecified overheads comprising the remainder (Herbohn *et al.*, 1997).

In relative terms, small cabinet-making firms are by far the largest users of rainforest cabinet timbers (Table 8.2). These firms also use relatively greater proportions of other high-value timbers such as other Australian hardwoods and imported tropical timbers (Herbohn *et al.* 1997; Petersen *et al.* 1997; Smorfitt *et al.*, 1997). The average proportion that rainforest cabinet timbers comprise of total wood products used by firms decreases dramatically with firm size, particularly in Townsville and Cairns (e.g. 32%, 18% and 6% for small, medium and large Townsville firms respectively). Conversely, medium and large firms in each region use a substantially higher proportion of composite wood products.

Table 8.1 Product mixes for cabinet-making firms in Cairns, Townsville and Brisbane

City	Firm size	Share of product type (%)		
		Furniture	Kitchens	Other
Cairns	Small	14	42	39
	Medium	11	71	18
	Large	16	59	25
Townsville	Small	46	40	14
	Medium	11	71	19
	Large	6	64	29
Brisbane	Small	24	51	25
	Medium	20	61	19
	Large	7	61	32

Source: Compiled from Herbohn *et al.* 1997, Smorfitt *et al.* (1997) and Peterson *et al.* (1997). Rounding errors cause some percentages across rows not to add to 100%.

The use of Australian rainforest cabinet timbers by North Queensland (Cairns and Townsville) firms is much greater than by Brisbane firms (Table 8.2). This is not surprising given that almost all supplies of these timbers are secured from North Queensland. For many uses, Australian rainforest cabinet timbers, other Australian hardwoods and imported tropical timbers are close substitutes. If these timbers are combined into a single category of 'cabinet timbers' then patterns of timber in Brisbane and North Queensland firms are much more closely aligned. There are a number of eucalypts that are commonly used in manufacture of kitchens. For this reason, the common eucalypt species (e.g. those sold under the trade name of Tasmanian oak) used by cabinet-makers have been categorised as 'cabinet timbers' although this is traditionally not the case.

In Townsville and Cairns most small firms surveyed use at least some RFCTs, with some using quite large proportions. Conversely, few medium and large firms use large amounts of RFCTs, with the averages for both of these categories being skewed by the relatively high use of these timbers by a small number of firms (see Herbohn *et al.* 1997 and Smorfitt *et al.* 1997 for further details). While RFCTs generally comprise a small proportion of the wood products used by medium and large firms, the absolute volumes may be large in particular cases. For example, for one Townsville firm with a turnover greater than $500,000, RFCTs comprised 40% of the total wood products used. In this case, the absolute quantities of RFCTs used would probably exceed the combined quantities of RFCTs used by all seven small firms in the sample.

Patterns of wood product use are almost certainly related to type of work undertaken by firms of different size. Rainforest cabinet timbers have traditionally been used for the construction of high quality furniture. This type of work makes up a high proportion of the turnover of small cabinet-making firms and explains their high usage of these timbers, along with imported tropical timbers with similar qualities.

Table 8.2 Use of rainforest cabinet timbers and composite wood products as inputs into products

Wood input	City	Fraction of total wood inputs used (%)		
		Small firms	Medium firms	Large firms
Rainforest timbers	Cairns	26	9	2
	Townsville	32	18	6
	Brisbane	7	4	3
Composite wood products	Cairns	50	64	84
	Townsville	33	49	64
	Brisbane	59	61	70
'Cabinet timbers'	Cairns	46	28	12
	Townsville	61	38	13
	Brisbane	27	25	14

Note: Cabinet timbers are the aggregate of 'rainforest timbers', 'imported tropical timbers' and 'other Australian hardwoods'.

Composite wood products such as chip-board and medium density fibre-board are the main materials used in the construction of modern kitchens. Solid wood inputs such as Tasmanian Oak (included in 'other Australian hardwoods'), and to a lesser extent rainforest timbers, are also commonly used in the exposed sections of high quality kitchens. Composite wood products are used with these timbers, in areas hidden from view. The relatively high usage patterns of Australian hardwoods by medium and small-sized firms strongly suggests that these firms specialise in the construction of higher quality kitchens. The small amounts of Australian hardwoods used by large firms, suggests that these firms mainly construct kitchens at the lower to middle of the quality and price range.

Just over a third of all cabinet-making firms kept no inventories of timber, with a further one third keeping less than $2m^3$ on hand (Herbohn *et al.*, 1997; Peterson *et al.*, 1997; Smorfitt *et al.*, 1997). There were however a number of notable exceptions, with nine firms in the sample holding between 10 and $65m^3$. Low inventory levels such as these strongly suggest that most firms only order materials when a job order has been placed. In Townsville where the level of use rainforest timber is highest, rainforest cabinet timbers comprised 78% of total timber inventories (Herbohn *et al.*, 1997). Furthermore, 13 of the 15 firms using RFCTs held inventories, and RFCTs comprised 90% or more of the inventories of eight of these firms. Conversely, only one of the nine firms not using RFCTs held any wood products as inventory, of only $0.3m^3$. This disproportionately high percentage of RFCTs in timber inventories compared with the levels of use suggests that firms have difficulty in obtaining these timbers on a regular basis, c.f. composite wood products are readily available from a number of suppliers on request. Similar patterns, although much less pronounced, were found in Cairns firms (Smorfitt *et al.*, 1997) though less so for Brisbane firms (Peterson *et al.*, 1997). The reason for this is almost certainly lower utilisation rates of rainforest timbers rather than greater availability reducing the need to hold stocks.

8.4 FACTORS AFFECTING THE CHOICE OF TIMBER INPUTS BY CABINET-MAKERS

Table 8.3 illustrates the degree of influence of cost, quality, suitability and customer requests on managers' choices of wood products. The high quality of rainforest cabinet timbers appears to be a factor in the decision of managers to use these timbers, with mean rating of quality as a factor in their decision to use them being significantly higher than for all other categories of timber ($P < 0.05$). The ratings for degree of influence of suitability of rainforest cabinet timbers on managers' decisions to use them were also significantly greater than for all other timbers except for Australian hardwoods and composite wood products. Few significant differences were found between various timber types with respect to customer request, with managers tending to attach a higher weight on this factor compared with the other three although these differences are not statistically significant.

Low cost is a factor in the decision of managers of cabinet-making firms to use composite wood products . The influence to low cost in the decision to use these products is significantly greater than for other categories of wood inputs. For these other categories, which are all solid wood products, there are no significant differences in the rating of low cost as a factor influencing choice. This suggests that once the decision is made to use solid wood inputs, cost is not a discriminating factor in choice of inputs. Solid wood products are considered expensive compared to composite wood products. If a customer requests the use of an expensive input, then the cabinet-making firm simply passes the cost of materials on to the customer. Costs of solid wood inputs such as RFCTs only become important to managers if the firm cannot pass the cost on to customers. While cost may not be an important factor in cabinet-makers decision *to use* RFCTs compared to other solid wood inputs, it is a major factor in managers deciding *not to use* rainforest timbers when costs cannot be passed on.

Customer requests for products to be made from rainforest timbers are received relatively infrequently in all three cities (Herbohn *et al.*, 1997; Peterson *et al.*, 1997; Smorfitt *et al.*, 1997). Significantly fewer customer requests for RFCTS are made to managers of Brisbane compared with managers in Cairns and Townsville. This indicates that customer request is an important factor in managers deciding on timber inputs.

Table 8.3 Influence of cost, quality, suitability and customer requests on managers' choice of wood products for use in the manufacture of kitchens and furniture (n = 24)

Wood product type	Low cost (lc)	Quality (q) (colour, grain)	Suitability (s)	Request by customer (r)	Significant differences (across rows)
Australian rainforest timbers (r)	2.5	4.1	4.0	4.2	lc < q,s,r
Australian hardwoods (ah)	2.9	3.7	3.8	4.2	lc < q,s,r
Australian softwoods (as)	3.1	3.4	3.5	3.9	c < r
Imported tropical timbers (it)	2.7	3.8	3.8	4.0	lc < q,s,r
Imported softwoods (is)	2.8	3.4	3.4	3.7	ns
Other timbers (ot)	2.9	3.6	3.7	3.8	ns
Composite products (c)	3.7	3.1	4.0	3.9	q < s
Summary of significant differences (down columns):					
	c>(r, ah, as, it, is, ot)	r>(ah, as, it, is, ot, c)	r>(as, it, is, ot); it>is	r>is, ot	

Note: Ratings are on a scale of 1 (very little influence) to 5 (very strong influence).

The frequency of requests from customers for products *not to be made* from rainforest cabinet timbers is also quite low, with 68% of managers reporting that they never receive such requests and a further 23% reporting that they seldom receive such requests. These results suggest that lack of customer demand is not due to environmental concerns. Some caution must be exercised when interpreting results because the survey addressed managers *not* customers. For example, customers who do not want products made from RFCTs for environmental reasons may simply request other products without making their views known to managers. Alternatively, customers with environmental concerns about the use of RFCTs may purchase non-timber products and hence have no contact with managers of cabinet-making firms.

Of the 130 managers surveyed, 65% had attempted to purchase Australian rainforest timbers in the past 12 months. There was no statistically significant difference ($P < 0.05$) between the three regions in proportion of cabinet-makers who had attempted to purchase rainforest cabinet timbers in the past year. Managers were asked to indicate the degree of difficulty that they had experienced in obtaining RFCTs in the appropriate quantities. A high proportion reported that they experienced difficulties either 'all the time' or 'often'. The problems were most pronounced in Townsville (63%) compared with Cairns and Brisbane (38% and 34% respectively). The relatively high inventories of RFCTs held by Townsville firms is also indicative of difficulties in obtaining these timbers on a regular basis.

Opinions in regard to the relative price of rainforest timbers were similar in the three cities (Herbohn *et al.*, 1997; Peterson *et al.*, 1997; Smorfitt *et al.*, 1997). The majority of managers (53%) considered RFCTs to be 'expensive' or 'very expensive' relative to other materials, while only 10% considered them to be 'inexpensive' or 'very inexpensive'. Perception of high price may be a reason for the low utilisation rates of Australian rainforest timbers, particularly by larger firms. Since large firms are predominantly manufacturing low cost 'mass produced' kitchens, made from less expensive materials such as composite wood products, it is unlikely that high input costs could be passed on to customers. Perceptions of high prices of RFCTs combined with an inability to pass these on to customers mean cabinet-making firms are unlikely to use these as inputs unless customers specifically request them.

8.5 FACTORS AFFECTING THE PURCHASE OF PRODUCTS MADE FROM RFCTs

Customers of cabinet-makers can be divided in one of three categories: those who buy, those who are interested but don't buy; and those who are not interested in buying. Responses by managers in regards to a series of questions relating to these three categories were similar in the three cities (Herbohn *et al.*, 1997; Smorfitt et al., 1997; Peterson *et al.*, 1997). In general, managers believed that customers purchase products made from rainforest cabinet timbers mainly because they prefer them to other products. Higher product quality and 'good value for money' were also considered important factors. Almost all managers believed that inability to afford products made from RFCTs was the main reason why customers were either *not interested* in buying these products or *interested in buying but not doing so*. Managers also believed that 'poor value for money' and 'preference for other timber products' were also factors contributing to the lack of demand for products made from RFCTs. Nearly half of the managers also considered that a customer's belief that rainforest cabinet timbers are no longer available is an important factor in why those customers interested in buying products made from RFCTs were not doing so. Hence, promotion of RFCTs to the general public may increase demand for such products.

The high level of agreement of managers with the statement that 'customers do not buy products made from RFCTs because they cannot afford them' indicates that a high final product price is probably a major constraint on demand. Part of the high cost of products made from RFCTs is undoubtedly associated with the high cost of RFCTs as inputs to the manufacturing process compared with other products such as composite wood products. In addition, the cost of labour associated with manufacturing products from RFCTs is likely to be substantially higher than for products made using other inputs. Because there is no secondary industry producing commonly used components such as doors and bench tops, all of these components must be manufactured on a one-off basis in-house. Mass-produced components commonly have a labour component of about 20% of total cost while one-off specialty items can have a labour component of up to 60% (McLennan, 1995).

In recent years much concern has been expressed by environmental groups about the effects of logging on tropical forests. Accordingly, public concerns about the effects of rainforest logging may influence the demand for products to be made out of RFCTs. About 30% of managers indicated that environmental concerns were a reason why people do not buy products made from RFCTs.

8.6 CONSUMER ATTITUDES TOWARDS PRODUCTS MADE FROM RFCTs

Surveys of consumers such as that undertaken by Smorfitt *et al.* (in press, b) provide important information for the marketing of RFCTs and products made from them. In that study, consumers were found to regard products made from RFCTs as being highly superior to those made from composite wood products. A majority also considered products made from RFCTs superior to products made from eucalypts, though about one third thought that the two were comparable in quality. A high proportion of people in the age between 25 and 54 had recently purchased products made from rainforest cabinet timbers (Table 8.4). While a high proportion of those over 55 had also purchased products made from RFCTs, nearly 50% of people in this category had not done so in the past five years. For the large proportion (56%) of people below 25 who had never purchased such products it is probably a 'stage of life' issue related to a lack of disposable income, combined with priorities for expenditure on other goods and services.

Members of the public who had purchased ARFCT products within the past five years were asked to indicate their level of agreement with seven statements concerning reasons for purchase. More than 50% of respondents agreed with the statements that 'realistically priced', 'good value for money', 'better quality', 'preference for RFCTs to other materials', 'liking for Australian timbers' and the 'desire to pass the product down as an heirloom' were reasons why they purchased RFCT products.

Table 8.4 Patterns of purchase of products made from RFCTs by the general public

Customer age	Fraction in age group (%)			
(years)	Last year	1–5 years	>5 years	Never
Under 25	23	18	3	56
25 – 34	21	31	14	34
35 – 54	23	33	23	21
55 or over	9	13	48	30

Source: Smorfitt *et al.* (in process, b); n = 249.

Members of the public who were interested in buying RFCT products but had not done so in the last five years were asked to indicate their level of agreement with a number of statements concerning reasons for not making a purchase. 'Poor value for money', 'poor quality', 'preference for other products' and 'destruction of rainforest' did not appear to be of major concern. Approximately 40% regarded these products as 'too expensive'. Fifty percent of respondents were undecided about whether these 'timbers were still available'.

A similar set of statements were presented to members of the public who had not purchased RFCT items and said they were not interested in doing so. Again 'cost', 'value for money', 'poor quality', 'preference for other products' and 'destruction of rainforest' did not appear to be of major concern. Again, 50% of respondents were undecided about whether these timber were still available. Sixty percent of respondents not interested in buying products made from RFCTs felt it was due to their 'stage of life'. This may include both the young who don't have the need for these products, or the resources to purchase them, and older respondents who are more likely to have fully furnished households.

8.7 CABINET-MAKERS ATTITUDES TOWARDS RFCTs AND PLANTING RECOMMENDATIONS

Cabinet-makers in Cairns, and to a lesser extent Brisbane, are familiar with a large number of rainforest cabinet species and rate many of these highly for use in kitchens and for furniture. Furthermore, there is a willingness to use these species even if they have not been used previously (Smorfitt *et al.* (in process, a). Table 8.5 presents opinions of Cairns and Brisbane cabinet-makers concerning what species should be planted to satisfy their future timber needs.

There is general agreement between both Cairns and Brisbane cabinet-makers about which rainforest and eucalyptus species they prefer. Although ranks of particular species differs slightly, five species are listed in the six most popular for both areas. The one exception is hoop pine, a native pine grown in large-scale plantations in south-east Queensland and to a much lesser extent on the Atherton Tableland in North Queensland. Hoop pine was ranked third by Brisbane cabinet-makers and only 19th by those in Cairns. This difference is possibly due to market penetration achieved for this species through a combination of ready availability, price competitiveness and promotion in the Brisbane market.

The results indicate that a number of rainforest and eucalyptus species, in particular Queensland maple, red cedar, northern silky oak, black walnut, Tasmanian oak and hoop pine, have sound market prospects. A comparison of the species recommended by cabinet-makers and those which have been planted under the CRRP in North Queensland reveals a notable disparity. Of the five most highly ranked species by Cairns cabinet-makers (all of which are native rainforest species), only Queensland Maple was planted to any extent (7.7% of CRRP planting up to 1997).

8.8 IMPLICATIONS FOR MARKETING OF NORTH QUEENSLAND RAINFOREST CABINET TIMBERS

Gresham (1995) suggested that the following factors (in order of priority) influence the purchase of timber by firms manufacturing wood products: colour, grain, volume availability, end-use range, price and physical properties.

In the current study, 'colour' and 'grain' were aggregated into the single category of 'timber quality'. The materials of choice for Queensland cabinet-makers are composite wood products and the factors which influence this choice are cost and suitability. Clearly, cabinet-makers are responding to market conditions where the high demand for indoor furniture and kitchens appears to be for a low-cost product made from a serviceable material. This evidence conflicts with Gresham's suggested order of priority for the factors influencing choice of timber product. Cabinet-makers, driven by customer demand, *are* willing to sacrifice quality for cost and serviceability.

Table 8.5 Comparison of Cairns and Brisbane cabinet-makers' top 15 species recommendations against CRRP plantings

Species	Cairns ranking	Brisbane ranking	Fraction of Cairns respondents (%)*	Fraction of Brisbane respondents (%)*	Fraction of total CRRP plantings (%)
Queensland Maple	1	4	83.9	60.0	7.7
Northern Silky Oak	2	5	82.1	51.4	0.5
Red Cedar	3	2	78.6	70.0	0.2
Queensland Walnut	4	6	75.0	51.4	**
Tasmanian Oak	5	1	67.9	77.1	***
Northern Silver Ash	6	7	60.7	47.1	1.8
Maple Silkwood	7	(17)	57.1	20.0	1.4
Qld. Silver Ash	8	8	51.8	45.7	1.4
Kauri Pine	9	14	51.8	28.6	6.0
Black Wattle	10	13	53.6	27.1	2.1
Black Bean	11	12	48.2	28.6	2.6
Red Silkwood	12	19	51.8	17.1	**
Satin Silky Oak	13	11	44.6	30.0	**
Red Siris	14	(20)	55.4	17.1	2.1
Hoop Pine	15	3	35.7	67.1	10.3
Rose Mahogany	(20)	9	28.6	48.6	**
Red Mahogany	(21)	10	25.0	38.6	12.7
White Beech	(23)	15	26.8	25.7	0.3

Source: Smorfitt *et al.* (in process, a)
Notes: * Timber was rated highly or very highly recommended
** Species either not in the planting list or less than 1,000 planted.
*** Multiple eucalyptus species.
Comparative rankings in parenthesis if out of the top 15 recommendations.

Only if composite wood products are excluded is there any measure of support for Gresham's suggested order of priority. For all other timbers, including

RFCTs, price is not a dominant factor influencing choice. Although 'quality' appears to have a strong influence, in relation to both indoor furniture and kitchens, as suggested by Gresham, 'customer request' is equally or more important. This pattern has implications for the use of *all* timber products in furniture manufacturing. Firstly, a majority of managers (53%) acknowledge that Australian rainforest timbers are 'expensive' or 'very expensive' (Herbohn *et al.*, 1997; Peterson *et al.*, 1997; Smorfitt *et al.*, 1997) and very few (10%) consider them to be 'inexpensive' or 'very inexpensive'. Secondly, a possible reason why the 'price' of an expensive product is irrelevant, from the manufacturer's viewpoint, is that quality timbers, such as RFCTs, are primarily used only when requested by the customer. In these circumstances, it is relatively easy for the manufacturer to pass the high timber cost on to the customer. Thirdly, customer requests to managers for products to be made from RFCTs are received relatively infrequently; 59% of managers in Brisbane and 36% in North Queensland reported that their customers seldom or never request products to be made from RFCTs.

Gresham has suggested that ready availability is an important factor in the choice of wood inputs. No direct evidence has been collected on whether availability of RFCTs has a major influence on the usage rates. However, in the Townsville and Cairns areas where the use of these timbers is greatest, managers reported higher levels of difficulty in obtaining rainforest cabinet timbers (Herbohn *et al.*, 1997; Peterson *et al.*, 1997; Smorfitt *et al.*, 1997). Furthermore, anecdotal evidence suggests that availability probably ranks with timber quality (colour and grain) and customer requests in order of importance. For example, imported Brazilian Oak, which is almost indistinguishable from locally sourced northern silky oak (*Cardwellia sublimis*) was recently selling for \$2,200/m^3. At the same time, *C. sublimis* was selling for \$1,800/m^3. Such a price differential for an almost identical product is surprising, especially given the high regard and long tradition of use associated with northern silky oak. One cabinet-maker, when queried on this, stated that firms using timbers such as these are willing to pay a premium for ready availability. Once again, this is because cost of timber inputs are passed onto customers; the customer pays the premium for ready availability (i.e. convenience for the cabinet-maker).

In Brisbane, problems of availability seemed less important with the majority of firms indicating they seldom or never had problems acquiring RFCTs (Peterson *et al.*, 1997). This is probably a function of much lower usage (associated with lower customer demand). Of greater concern in the Brisbane market is that many firms (30%) indicated that they had not attempted to acquire RFCTs over the previous year because *they do not use the material*. Hence, it would appear then that even if ease of access to rainforest cabinet timbers was increased, this would not necessarily result in increased use.

In terms of price and quality, imported tropical timbers and Australian hardwoods can be regarded as substitutes for RFCTs. There is anecdotal evidence to suggest that both of these categories of timber inputs are readily available to cabinet-makers on demand and that this is a major reason for their use in preference to RFCTs. Therefore, increasing the ease by which RFCTs can be acquired would make them more competitive with their direct substitutes. However, the current use of Australian hardwoods and tropical timbers by most firms is not great, particularly for large firms, so there is only limited scope for substitution in absolute quantities. Also, RFCTs are not perfect substitutes, particularly for Australian hardwoods. The latter are a more convenient timber input for cabinet-makers because of the availability of pre-manufactured products (e.g. bench-tops, cupboard doors) not currently available in RFCTs.

If efforts are to be made to increase the utilisation of RFCTs and Australian hardwoods, other mechanisms apart from price and availability must be considered. One such mechanism is by increasing customer demand for products to be made from RFCTs through better marketing. Since RFCTs are held in high regard by cabinet-makers in terms of quality and suitability for use in kitchens and furniture, there is potential to market RFCTs to the public as being a superior product (i.e. diamond timbers). This potential will not be achieved unless the industry adopts better marketing strategies.

Sinclair (1992) identified a number of factors impacting on the demand for furniture. These factors include personal disposable income, interest rates (and thus the cost of financing purchases), population geographic movement (with resulting replacement of furniture) and demographics of the household and household formation. A study by Epperson and Wacker (1989) identified the 25 to 54 year age group as having the highest expenditure on furniture, associated with household formation.

Promotional campaigns can be used to increase market share by increasing the public's awareness of a product. One approach would be to reduce prices, but this does not appear to be feasible under current circumstances where higher stumpage prices for RFCTs are desired. An alternative is to market tropical timbers as 'diamond timbers' as suggested by Johnson and Sarre (1995) from the International Tropical Timber Organisation Secretariat. Diamond timbers are those suitable for decorative use, e.g. top-of-the-range furniture and feature panelling, and timbers for restoration of antiques and old buildings, musical instruments and hand-made arts and crafts items. This principle could in turn be extended to products made from these timbers. Johnson and Sarre have suggested that there is great potential for high prices to be obtained for these timbers because of their aesthetically pleasing characteristics and inherent physical stability. Many Australian rainforest species exhibit the characteristics of 'diamond' timbers and some, such as red cedar, already command high prices.

Before rainforest cabinet timbers can be promoted and sold as 'diamond' timbers, an understanding of the final markets must be achieved (Johnson and Sarre, 1995), particularly in respect to the factors affecting the choice of timber inputs. The research reported in this chapter is an important first stage in providing such an understanding.

A central marketing organisation for rainforest cabinet timbers could aggregate the current fragmented supplies of these timbers, and substantially increase the ease with which cabinet-makers could gain access to them. The studies reported by Herbohn *et al.* (1997), Peterson *et al.* (1997) and Smorfitt *et al.* (1997) suggest that making these timbers readily available to cabinet-makers is probably a necessary precursor to a marketing campaign designed to raise the profile of these timbers with the general public. It may be an easy task to persuade consumers that RFCTs are a high quality material, superior in many respects to other materials. It may also be possible to improve the availability of supply of RFCTs although this will not be as easy. The difficulty will be to persuade consumers that it is worth paying a premium for this quality. The task will be made even more difficult because these substitute materials have already gained market acceptance.

8.9 SUMMARY

Australian rainforest cabinet timbers are highly regarded by both cabinet-makers and the general public. Small cabinet-making firms are the largest users of RFCTs; usage by medium and large firms is quite low, with their product of choice being composite wood products. Four factors appear important in the choice of inputs by cabinet-makers – quality, availability, customer requests and cost. Cost is only important when it cannot be passed on to customers. Where cost is a factor, the product of choice is composite wood products. Cabinet-makers are willing to pay a premium for availability and will avoid products that are not readily available unless customers specifically request their use. Many cabinet-makers have difficulty obtaining regular supplies of RFCTs and in many cases have been turning to other products. In the southern markets, the data were even more concerning – large numbers of cabinet-makers, despite having a very high regard for RFCTs, are simply not trying to obtain them. The data reported in this chapter gives support for moves by local growers to form co-operatives to market timber. These could help overcome some of the availability problems and thus lead to greater demand. However, to increase demand by cabinet-makers, customer numbers must first be increased.

Cabinet-makers will probably pay higher prices for the RFCTs as long as they can pass this on to customers. Another finding of this chapter is that RFCT species being planted in North Queensland are not a close match with those expected to be in greatest demand in the future.

REFERENCES

Adam, P. (1994), *Australian Rainforests*, Oxford University Press, Oxford.

Carron, L.T. (1993), 'Changing the nature of Federal-State relations in forestry, Australia's ever-changing forests II', in J. Dargavel and S. Feary (eds), *Proceedings of the Second National Conference on Australian Forest History*, Canberra: Centre for Resource and Environmental Studies, Australian National University.

Epperson, W. and Wacker, B. (1988), *1988 Home Furnishings Compendium*, Wheat First Securities, Richmond, VA.

Gresham, G. (1995), 'In quest of the unknown: marketing is the key to increasing the use of lesser known species in trade', *ITTO Tropical Forest Update*, **5**(2), pp. 3–5.

Herbohn, J.L., Smorfitt, D.B., Peterson, R. and Harrison, S.R. (1997), *Cabinet-making firms and the demand for rainforest cabinet timbers: (I) Townsville survey*, Research Report Series No. 7, Townsville: James Cook University of North Queensland.

Johnson, S. and Sarre, A. (1995), 'Economics and natural tropical forest management: An overview of the economic hurdles facing natural tropical forest', *ITTO Tropical Forest Update*, **5**(4), pp. 3–6.

McLennan, J. (1995), 'At what price profit', *Australian Wood Review*, No. 9, Summer Edition, pp. 27–30.

Moore, P. (1992), Rainforest timbers and cabinetwoods: marketing considerations, *QFS Field Day, Mt Mee.*

Peterson, R., Smorfitt, D.B., Herbohn, J.L. and Harrison, S.R. (1997), *Cabinet-making firms and the demand for rainforest cabinet timbers: (III) Brisbane survey*, Research Report Series No. 9, Townsville: James Cook University of North Queensland.

Smorfitt, D.B., Herbohn, J.L., Peterson, R. and Harrison, S.R. (1997), *Cabinet-making firms and the demand for rainforest cabinet timbers: (II) Cairns survey*, Research Report Series No. 8, Townsville: James Cook University of North Queensland.

Sinclair, S. (1992), *Forest Products Marketing*, NewYork: McGraw-Hill.

Smorfitt, D.B., Herbohn, J.L., Peterson, R. and Harrison, S.R. (in process, a), Rainforest cabinet timber species recommended by cabinet-making in order to satisfy their future timber needs: A possible guide for future planting and research.

Smorfitt, D.B., Herbohn, J.L., Peterson, R. and Harrison, S.R. (in process, b), Public perceptions about the availability and utilisation of rainforest cabinet timbers and the implications for marketing.

9. Developing Export Markets for Australian Forest Products

M.A. Cox and M. J. Quayle

This chapter examines issues surrounding the export of forest products from Australia, with an emphasis on forest products of sub-tropical Queensland. There is a need for greater export awareness due to a projected surplus in Australian timber production in the next decade. Value-added products appear to have the best potential for export, and co-operative marketing may be a key factor to facilitate export growth. The Asia-Pacific region appears to offer the best prospects for an export programme.

9.1. INTRODUCTION

The Australian domestic supply of logs is approaching self-sufficiency, and a timber surplus is expected by 2010. The ability of some South-East Asian producers to supply forest products has been reduced by government fiat or exhaustion of resources, and many global studies have predicted a timber shortfall in the Pacific region in the next 20 years. This chapter argues that a greater emphasis on exporting forest products is the best response by Australian producers to increasing domestic supply. It explores issues in the export market for timber and examines the potential of some finished products and markets for exports, with an emphasis on the Queensland timber industry. Issues of supply, potential for and development of trade, and impediments to trade are examined.

9.2. FACTORS AFFECTING TIMBER SUPPLY, CONSUMPTION AND EXPORTS

The Australian share of the world's forest area is only about 2%. However, low population relative to land area and advanced forest management techniques places the country in a strong position to produce surplus forest products for export. Out of 157 M ha of general forest area, about 1.2 M ha is State or private plantation forest (ABARE 2000). Native forests provide considerable woodchip exports. Plantations consist mainly of conifer softwood species, including radiata pine *(Pinus radiata)*, slash pine *(P. elliottii)*, caribbean pine (*P. caribaea)* and hoop pine (*Araucaria cunninghamii)*. In 1998, an estimated 948,000ha was plantation softwood and 287,000ha plantation hardwood. This represented a growth of nearly 19% in the three years from 1995. The farm forestry sector is adding to this sustainable supply through various planting schemes and alliances with industrial scale forestry. Tasmania and Western Australia have formalised schemes that allow small-scale producers to benefit from industrial processing economies and security of markets.

Queensland is the only major producer in tropical Australia, having about one sixth of the plantation area (200,000ha), of which 90% is government owned. Most is exotic conifers – caribbean pine, slash pine and hybrids; about 25% is the native hoop pine. A growing supply of lesser known species is becoming available from farm forestry. In Queensland, DPI Forestry (1998) has forecast plantation supply to peak in 2002 at about 2.2 M m^3 and a sustainable yield of 1.8 M m^3 to be achieved by 2010.

National policies have been adopted to increase supply, including *Plantations for Australia: the 2020 Vision* (Ministerial Council on Forestry, Fisheries and Aquaculture 1997) which aims to treble the Australian plantation area by 2020. Success of this proposal would mean a plantation area in excess of 3 M ha, with up to 40 M m^3 of timber harvested annually, creating a greater need for export outlets of forest products.

In contrast to plantation growth, total timber supply from native forests has been reduced by the Regional Forestry Agreements (RFAs) between the Federal and State governments initiated by the National Forest Policy Statement in 1992 (Commonwealth of Australia, 1992). The RFAs are designed to provide security of supply for the industry while ensuring that representative areas of various ecosystems are protected. Security of supply is a key requirement for investment, for processors in both the domestic and export sectors.

The domestic consumption of forest products within a country is likely to change over time, and demand differs between nations at different stages of development. Such changes in demand in South-East Asia provide export opportunities for Australia. Table 9.1 presents international annual per capita industrial wood usage in the 1990s. Differences in demand between developed and developing countries support an argument by Neilson (1998) for the potential for increased exports from the former and increased consumption of timber products and imports in the latter.

Demand for forest-based products depends on many factors. These include availability of substitute building products, changing price ratios of competitive timbers and consequent affordability of alternative products, changes in environmental attitudes, availability of traditional timbers, and fashion changes that indicate growth in niche market demand (such as the growth in timber flooring markets). Farm forestry has the potential to grow more diverse timber species and to differentiate its product in an export market and find demand in countries with sophisticated consumers looking for niche products. It has many practitioners who emphasise sustainable use of their forest asset, which is necessary to satisfy certification and labelling requirements of tropical timbers in some markets.

Table 9.1: Annual per capita industrial wood use in selected markets

Country	Per capita usage (m^3)	Country	Per capita usage (m^3)
Sweden	3.35	Malaysia	0.66
United States	1.78	South America	0.33
New Zealand	1.74	Indonesia	0.11
Japan	0.94	China	0.10
Australia	0.86	India	0.03
Singapore	0.78		

Source: Neilson (1998).

9.3 DEVELOPMENT OF TIMBER EXPORTS

Figure 9.1 indicates the relative changes in domestic production, imports and exports of coniferous and broadleafed timber in Australia over the period 1953–97. Since the late 1970s, exports and coniferous production have grown, but reliance on imports has also increased. One aim of trade is to increase the value of exports at a rate faster than the value of imports and an accepted key to this is to increase the amount of value added to timber products. An example of this pattern is the growth in Queensland's export trade for sawn timber through the 1990s (Figure 9.2). While there has been considerable volatility, recent statistics indicate a more rapid increase in the per unit value of exports than the per unit value of imports.

Development of trade is desirable where there is underutilisation of the existing resource in the domestic market. The surplus of industrial timber in the southern states of Australia and the planned growth of forest estate under the 2020 Vision Statement point to this occurring. Australian tropical timber processors are well placed to take advantage of recent events in international supply of forest products. The United States trade market since 1993 has been affected by the environmental restrictions placed on export of logs.

Exploitation by Asian and Pacific rim countries of their forest estate and consequent timber shortages have also led to opportunities for timber trade. Malaysia provides an example of a trade competitor that supplied tropical sawlogs and sawnwood to the international market at an unsustainable rate. Forest depletion in Malaysia was estimated at 700,000ha/year in the early 1990s. Logging bans were introduced, and domestic timber companies sought supplies from other Pacific countries, including Papua New Guinea, Indonesia and the Philippines. These countries subsequently established policies to manage their forest resources, under the direction of the International Tropical Timber Organisation (ITTO), and introduced their own logging bans. The reduction in ability to supply timber to the international market from these actions has created opportunities for the supply of tropical timber from Australia.

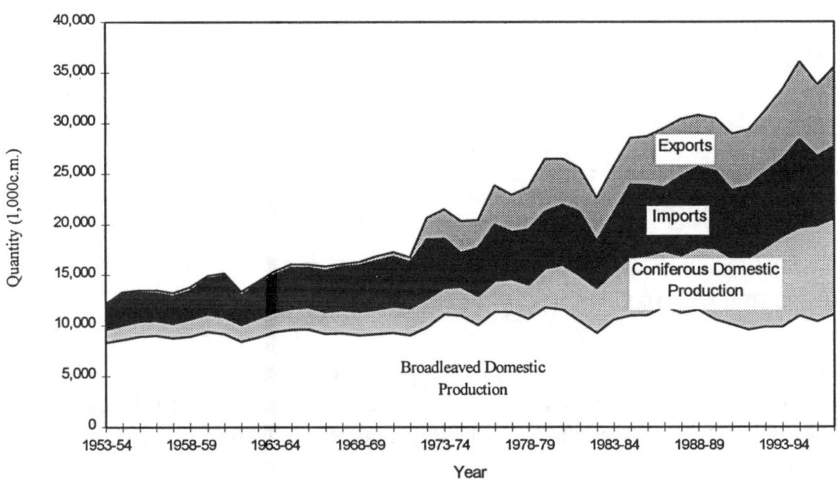

Figure 9.1 Relative changes in components of Australian timber production and trade

Source: ABARE (2000).

Developing trade provides an opportunity to insulate the domestic market against downturns in the domestic business cycle. The influence of demand cycles of the housing industry on Australian prices for timber provides an example of how domestic timber producers have been captured and influenced by their domestic market. Where markets have been opened up to international trade influences, such as in New Zealand, log prices have accelerated past their domestic pre-trade levels. The prices received for

Australian tropical forest products from small-scale forestry would benefit in a similar way from a much wider market for their product.

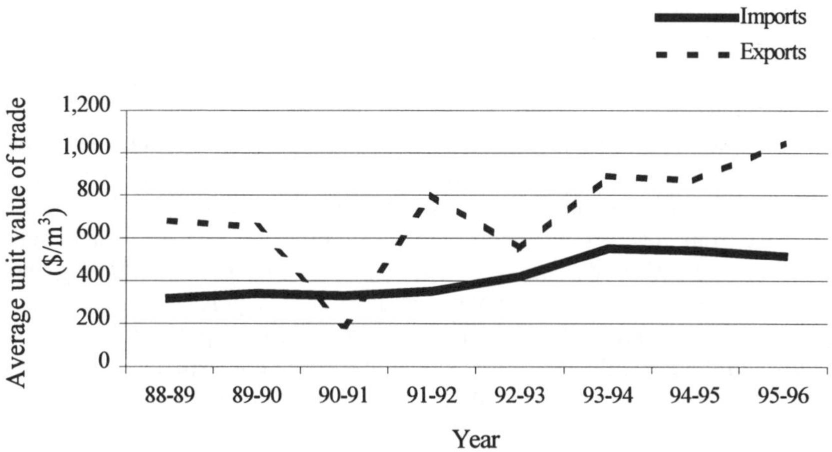

Figure 9.2 Value of Queensland sawnwood exports trade

Source: DPI Forestry (1998).

Forestry in Australia generally has a commitment to sustainable use of the forest estate, which makes exporters well placed to take advantage of moves for certification and labelling (C & L) of forest products. C & L is an area whereby international bodies such as the World Trade Organisation (WTO), Food and Agricultural Organisation (FAO) and ITTO hope to support sustainable development of tropical forests and allow their use 'by producers as tools to promote market access and enhance trade opportunities' (Liu 1996, p. 1). The aim is to use market incentives to encourage forest management and to facilitate trade in products derived from sustainably managed forests. The Australian forest industry generally has committed itself to this development for business and marketing reasons. Farm foresters with a strong commitment to the environment find that C & L fits well with their approach to forestry.

In Queensland, both native forest and plantation land are under the control of the Department of Natural Resources (DNR), but plantations are managed by DPI Forestry. The DNR ensures that the requirements of the National Forest Policy Statement and Regional Forest Agreements are adhered to. C & L for Queensland is attainable without dislocation of the forestry industry. This eases the path for exporters in major export markets where sustainable forest management is becoming important for consumers, such as in Europe and the USA.

Because of the relatively small size of the Australian population, it is difficult for timber processing facilities to achieve the lower unit costs of large traditional suppliers such as Canada and the USA, if producing only for the domestic market. Exports provide access to larger markets that can allow economies of scale and hence lower unit costs for producers.

Small-scale foresters in some areas of Australia have created alliances with industrial forestry for processing log output to benefit from economies of scale. There is potential for 'downstream' value-added products in the industry from these arrangements, and access to export markets. The benefit from an economic and marketing perspective would come from finding new markets, developing new products and finding export niches for their products that would allow them to increase their output and especially withstand domestic demand fluctuations in volatile domestic niche markets. As an example, relatively small hoop pine processors in tropical North Queensland have used finger jointing of hoop pine for mouldings and laminated board product to increase product value and develop products for new markets in Japan and southern USA.

9.4 IMPEDIMENTS TO EXPORTING FOREST PRODUCTS FROM AUSTRALIA

The development of trade is a desirable aim for farm forestry but there are difficulties faced by both small and large-scale producers in entering into or expanding export trade:

(i) *Supply uncertainty*. A major historical impediment to exporting from Australia has been the uncertainty that processors experienced about supply of timber. Uncertainty can influence expectations and investment plans, and create a reluctance to enter long-term overseas export arrangements. Long-term supply arrangements from native forests have been discouraged by state forest agencies that are dealing with the changes created by their corporatisation and privatisation. The traditional security of supply in Queensland where processors were provided with 20-year contracts in the 1980s is no longer possible. The issues of conservation and environmental management have also intervened to discourage long-term contracts. The 1992 National Forest Policy Statement has recognized the need to improve resource security. Plantation timber resources have grown quickly since the 1970s from both private and government sources to help meet supply objectives. Small-scale forestry by contrast is not well coordinated in its approach to planting and consequently supply is less stable.

(ii) *Non-price issues.* In an export market the non-price issues of quality, brand name, reliable supply, consistency of product and efficient delivery mechanisms are crucial. The Furniture Manufacturers of Australia Association (FMAA) point to the need for economies of volume to allow for quality improvement (Industry Commission 1993), and this illustrates the interdependent nature of the export decision for forest product sellers. The chain of forest management, production, marketing and exporting impinges upon the viability of the decision to export. This may imply a greater need for vertical integration or partnership arrangements to facilitate export programs (Cameron and Biggart, 1998).

(iii) *Lack of export experience.* Lack of experience of Australian producers in export markets is an impediment to international competitiveness. The small-scale forestry sector could benefit from emulating the marketing approaches pursued by industrial timber processors. The hoop pine processors of Queensland (Araucaria Australia Ltd) provide an example of how to overcome this lack of experience. The araucaria industry processors created a business entity to act as a focus for their group and sought funding from government and industry to develop a unified export marketing approach. Many of the processors were inexperienced in export marketing so a series of seminars were held and a trade tour of potential markets in China and Japan was conducted to give the industry members greater knowledge and exposure in these areas. Research and development of new products to meet the needs of the export market has been a continuing commitment of the group. Single points of contact and more standardised packaging and pricing are other issues that have to be addressed in export markets. For relatively small suppliers in a world market a cooperative approach makes good economic and marketing sense.

(iv) *Lack of species recognition.* Australian hardwoods and rainforest species are not widely recognized overseas, and small-scale foresters often plant these species. Queensland industrial plantations comprise mostly of species well recognised overseas and are thus readily acceptable to international markets. In contrast, the native species most widely grown – hoop pine – is not traded internationally in large quantities and can suffer by low recognition and hence lower prices even though it has superior qualities to other conifers. In the same way, a secure name is yet to be established for eucalypt hardwoods in overseas markets, although marketing groups in Victoria and Queensland are now promoting eucalypts overseas.

(v) *Need for government support.* Australian government assistance has generally been in the form of import tariffs which in the 1980s were 10% on sawn timber and up to 40% on plywood. Phased reductions bought this to 5% on imported wood and wood products in 1996. Only domestic producers who can be internationally competitive are likely to survive in the long run if commitments to lower protection are pursued. Government support for the domestic industry has not been as generous as our near competitors, e.g. investment tax allowances, tax holidays while businesses are established, export market development support and training are provided to Malaysian firms.

(vi) *High transport costs*. Australia including Queensland has suffered from two major infrastructure impediments to exporting timber products, viz. high cost of handling on Australian wharves and unavailability of low-cost shipping. (McComb, 1998) estimated that the freight disability from the wharves – from extra port charges, lower loading and stowage rates and higher stevedoring costs – added \$5-\$8/m^3 to the cost of log exports from Australia in the 1990s. The international shipping freight cycle also disadvantages Australia. In boom times for international trade the Australian market is charged a premium on freight since Australia is not seen as a favourable loading country while other cargoes are to be had from more efficient ports and ones that have profitable backloads. This premium can add \$1 to \$3 per tonne of cargo to the cost of transport.

(vii) *Long intrastate distances*. Tropical timber products from regional centres such as North Queensland face severe domestic transport cost constraints. To cover transport costs, it is necessary to produce value-added products such as ply, veneer, laminated board, finger-jointed mouldings, specific rare timber logs or furniture products. Small-scale forest producers have to consider the importance of processing of timber into more valuable products to outweigh this impediment. A Queensland illustration of this is the high-quality timber resources of Cape York, which are presently a non-economic resource, with lack of roads and long sea distances to domestic markets.

9.5 TIMBER PRODUCTS FOR EXPORT MARKETS

Australian trade in forest products has been concentrated on the primary outputs of forests. The potentially more rewarding export trade in value-added forest products has been neglected by comparison. This situation is changing as investment in processing capacity and new timber products emerge. The forest products that are being traded or have the potential to be traded from Australian forest timbers are discussed below.

(i) *Processed logs and woodchips*. Australian trade has been primarily in the export of processed logs, with the dominant value in pulpwood or woodchip products where there has there been a consistent export surplus. Reconstituted board products, sawntimber and paper and cardboard have been less prominent. Sawntimber exports from Australia are almost equally softwood and hardwood, and the two main markets are Japan and the United Kingdom.

(ii) *Panel products.* Wood panels produced in Australia include medium density fibreboard (MDF), particleboard, softboard, hardboard and plywood. Imports of these products in the past have encouraged the establishment of domestic MDF plants by local and NZ companies. This import substitution in the domestic market has increased domestic supply. MDF was first produced in 1982 in Australia and by 1996 Pettingill (1998) estimated output was 550,000m^3 accounting for over 30% of wood-based panel consumption in Australia. In 1995–96, MDF panel exports were 896,650m^3, and between 1997–98 to 1998–99 total MDF exports increased from 1.39 to 2.06 M m^3 even as domestic production declined slightly. Domestic MDF consumption is predicted to rise 50% over the next 20 years (Connell *et al.*, 2000). New markets in China were opened up in 1999 and by the September quarter exports to China had reached 12,000m^3 (ABARE, 2000). Currently, Australia's main export markets for wood-based panels are New Zealand and Asia. A potential market exists for small-scale forestry in the use of tropical forest timber veneers and hardwood timbers in export panels. BIS Shrapnel identified the growing use of MDF as a replacement for plywood, particleboard and sawn timber and its penetration into new markets in Asia and South America (Neufeld and Pettingill, 1997). MDF has uses in shop fitouts, furniture manufacturing, stair treads, kitchen cabinets and cupboard doors and mouldings.

(iii) *Furniture timber industry*. The decline in international availability of slower growing temperate hardwoods such as oak, beech and walnut and the tropical species of mahogany, teak and meranti due to resource depletion and conservation measures provides market opportunities for Australian native timbers. Selective and sustainable logging of privately-owned native forests or from logging state forests and government eucalypt plantations allowed under Regional Forest Agreements could be used as a basis for supply of high-quality hardwood timber to the furniture industry. A major domestic consumer market change in approach to furniture timber is the growing acceptance of Natural Feature (NF) hardwoods. The provision of NF standards, research, technical support and support for export provide a future direction for farm forest timber. As an illustration of its success in the domestic market, the top selling bedroom suite in Australia is made from natural feature messmate (*Eucalyptus obliqua*) (Gooding, 1999).

(iv) *Engineered wood products*. World-wide, the availability of large sawlogs from native old-growth forests has declined; one response has been to adopt engineered wood products to provide alternatives for structural timber. World timber markets have already adjusted to the use of softwoods rather than hardwoods for structural sawn timber, because of their plantation availability, price and sustainability of production. The development of a laminated veneer lumber (LVL) market in Australia has paralleled the growth of softwood plantation markets in Europe and the USA where LVL has replaced steel beams because of cost considerations. The LVL process aligns the grain of timber sections in the same direction and then glues them together under pressure. It can create up to 160mm deep panels which can be sawn as if they were solid lumber. Both softwood and hardwood can be used to create LVL products, and strength can be specified and guaranteed. Wooden I beams can be used as structural members in place of steel and can lower cost and weight in multiple floor domestic and commercial building structures. The Japanese market also makes use of LVL in non-structural applications such as stair balustrades, treads and risers, window frames and doors.

(v) *Housing products*. Furniture, flooring and windows are value-added timber products that have export potential. In Queensland's case the use of hardwoods such as blackbutt (*Eucalyptus pilularis)* and spotted gum (*Corumbia maculata)* for flooring and decking because of their durability and strength are potential export products in a market where hardwood products are in short supply. For softwoods, the state's cypress (*Callitris glaucophylla)* forests have sustainable yields of 120,000m^3 per annum with private harvests stabilised at around 60,000m^3 and these are available for export markets. Cypress has excellent properties for flooring, being termite resistant and useful for framing, fencing and structural uses. Competition with lower priced softwood substitutes in flooring has encouraged greater efforts in quality control such as matching, planing and drying techniques to differentiate the product. Cypress could compete well against rubberwood which is used in Japan, the US and Europe for flooring. Rubberwood twists in lengths over 1.2m so in flooring is mostly confined to parquetry and not sold in lengths as strip flooring.

(v) *Factory built dwellings*. The European and USA markets are large purchasers of ready-made timber housing products, in finished form or in ready-to-assemble kits produced in a factory assembly line. As an example of the potential size of the market, factory-built single family homes accounted for 25% of homes completed in the US in 1990, and 90% in Sweden (CINTRAFOR 2000). Population growth and forecast housing requirements in Asia indicate a potential market where Australian product could compete, given our proximity to these markets.

9.6 MARKETING APPROACHES TO THE EXPORT OF TIMBER

Entry into new markets often requires assistance from either government or industry associations. Under the 'level playing field' approach of recent Australian governments, assistance has been difficult to obtain, except for support in some regional areas for political purposes and usually for the benefit of domestic rather than overseas trade. Few Australian processors or manufacturers of timber product have the funds or knowledge to allow them to open major new markets by themselves, and any chance of trade influence based on size or value would be small.

The Queensland Timber Board provides an example of an industry association that is attempting to support the marketing of a specific forest product – hoop pine (*Araucaria cunninghamii*) sawn timber – into the major markets of Japan where the characteristics of close grain, light appearance and good working characteristics are highly valued. In Victoria the Australwood group has successfully marketed hardwoods under the 'Golden Forest' label. The marketing lesson for potential exporters is that there is strength in a combined effort and one that has the support of industry bodies and government.

9.7 EXPORT MARKETS AVAILABLE TO THE AUSTRALIAN FOREST INDUSTRY

The traditional and dominant market for traded forest products from Australia has been Japan. This market is one that attracts many traders in forest products because of its size and the reliance of the Japanese on imported pulp, logs and other timber products. The Japanese have also established ownership or control in timber processing companies and plantations around the Pacific rim to ensure some control over their supplies of raw timber and pulp.

The Japanese market illustrates some features required to provide an export market for timber processors or value-adding producers who wish to export. If producers can successfully export to Japan then other regional markets will most likely follow. The market has large size, population and demand for timber products, and proximity to Australia. There is demand for a range of products from logs to highly value-added output such as engineered wood products and furniture. Not all regional markets have the same range of market characteristics as Japan but they do share in the expected shortfall of supply predicted for the Pacific region generally.

Global outlook studies for forest product demand have generally concluded there will be an excess of demand for timber soon after 2000 (FAO 1997). Some forecasting models predict demand in Asia will outstrip supply (Apsey and Reed, 1995; Nilsson, 1996). Examination of markets for tropical and non-tropical timber forest products have usually been confined to reports by consultants such as D. A. Neilson and Associates (DANA) and the Centre for International Trade in Forest Products, Washington University (CINTRAFOR), University centres, and institutions such as the FAO and ITTO. A general consensus is that the Pacific region will experience undersupply until at least 2020. Although timber substitution and new plantations are likely, export opportunities plainly exist.

Australia may be able to exploit its advantage of proximity to these markets and differentiated timber species, with low processed timber product such as logs and pulp, and more value-added exports. Other producers such as New Zealand and Chile have the same opportunities and similar timber product, which could imply reduced prices in the future in a competitive market especially for radiata pine. Chile, which has concentrated on the US market and Europe up to now, has large radiata plantations like New Zealand, and Australian exporters will have to compete on the basis of quality of product and certainty of supply. Differentiating output may be the appropriate marketing approach.

Available markets in the region are Japan, Taiwan, China, Korea, Thailand, Indonesia, Philippines, Singapore and Vietnam. These countries belong to either ASEAN or APEC, both of which have committed members to reductions in tariffs to promote trade. Tariffs have in the past prevented large-scale timber exports, but these have been decreased as timber supplies in these countries have declined due to excess harvesting and political actions to conserve their resource. Tariffs on remanufactured timber goods such as panels (15–20%) and furniture (30–60%) have been typical on traded forest products. These levels are declining as tariff reduction agreements are met. ASEAN countries are committed to reducing tariffs to 5% by 2008 for goods which meet member rules of origin.

In addition to specific regional agreements, signatories to the 1994 GATT Uruguay round have agreed to eliminate tariffs on furniture products completely in the next decade. This means major importers of furniture product such as Japan, the USA and members of the European Union have agreed to open their markets to these products. Australian hardwood timbers have some unique features as well as being managed for sustainable yields, which may help gain entry to these markets. Although the competitive strengths of Australian hardwoods have not rested in design of furniture and timber products in the past, the development of this area could provide the outlet for a profitable export industry and align with unique flooring and veneer products. The Australwood group of Victoria has successfully marketed Victorian hardwoods to China and Japan for flooring and furniture.

A co-operative approach to trade between processors and growers might include pooling of research into timber types, market evaluations, export transport arrangements and even central processing facilities. Team efforts at market entry seem more likely to be successful than individual attempts. This implies a role for governments, timber boards and industry leaders, and regional alliances of small-scale forestry with industrial processors.

9.8 CONCLUSIONS

There are a range of arguments and issues that can be used to support a move to exporting, central to which is a pursuit of efficiency. The Australian timber industry will face the need for greater export effort. The nature of the industry is likely to change as increased production in Australia and undersupply in the Pacific region encourage greater exports. Strategic advantage will arise from moving into value-added trade in forest products, particularly panel products such as MDF, engineered wood products such as laminated lumber and furniture and niche markets for tropical timber forest products. The dominant export markets will remain regional as reductions in tariffs open up these markets. This may take the form of greater involvement in the Japanese market and more trade with Taiwan, China, Indonesia and the Philippines. Markets further afield could be entered with higher value-added products such as furniture and flooring timber. Small-scale forestry in the tropics has to position itself to benefit from the potential for growth in trade that the development of export markets provides.

REFERENCES

ABARE (Australian Bureau of Agricultural and Resource Economics) (1999), *Australian Commodity Statistics*, ACS 99, Canberra: ABARE.

ABARE (Australian Bureau of Agricultural and Resource Economics) (2000), *Australian Forest Product Statistics*, September quarter 1999, Canberra: ABARE.

Apsey, T.M. and Reed, F.L.C. (1996), 'World timber resources outlook: current perceptions with implications for policy and practice', *Commonwealth Forestry Review*, **75**(2), pp. 155–159.

Cameron, J. and Biggart, R. 'Optimisation of the Pine stump to Market value chain', Paper presented at the National Australian Forest Industries (NAFI) Conference, Sydney, April 1998.

Centre for International Trade in Forestry (CINTRAFOR) (2000), 'Factory built construction: domestic and export developments,' http://www.cintrafor.org/FS09.htm , accessed Feb. 2000.

Commonwealth of Australia (1992), *National Forest Policy Statement: A New Focus for Australia's Forests,* Canberra: Australian Government Printing Service.

Connell, P., Spencer, R., Andrzejewski, K., Sledge, P. and Whitfield, D. (2000), 'Prospects for non-sawlog wood in Australia', in ABARE (ed.), *Australian Forest Product Statistics*, September quarter 1999, Canberra: ABARE.

Department of Primary Industries – Forestry, (DPI Forestry) (1998), *An Overview of the Queensland Forest Industry*, Information Series QI 98015, Brisbane: DPI Forestry.

Food and Agricultural Organisation (FAO) (1997), *State of the Worlds Forests*, Rome: United Nations.

Gooding, G. (1999), 'Market opportunities for high value timber products', in ABARE (ed.), Outlook 99, Proceedings of the National Agricultural and Resources Outlook Conference, Canberra, 17–18 March, Vol. 1, Commodity Markets and Resource Management, ABARE, Canberra, pp. 251-258.

Industry Commission (1993), *Adding Further Value to Australia's Forest Products,* Report No. 32, Canberra: Industry Commission.

Liu, V. (1996), 'Trade aspects of certification and labelling', in Session 6 *International Conference on Certification and Labelling of Products from Sustainably Managed Forests*, Sydney: DPIE, URL access http://www.affa.gov.au/dpie/conference/iocal/papers/6over2.html, accessed Feb 2000.

McComb, M. (1998), 'Maximising Australia's export potential and case study of Radiata Exports Pty Ltd', Paper presented at the National Australian Forest Industries (NAFI) Conference, Sydney.

Ministerial Council on Forestry, Fisheries and Aquaculture (1997), *Plantations for Australia: The 2020 Vision*, Canberra.

Neilson, D. (1998), 'Outlook for wood suppliers: Can Australia grow wood for the international market?', paper presented at the Australian Timber and Forestry Conference, Sydney.

Neufeld, B. and Pettingill, J. (1997), *Medium Density Fibreboard in the Pacific Rim, 1997–200*, Sydney: BIS Shrapnel Forestry Group Study.

Nilsson, S. (1996), *Do we have enough forests*, Laxenburg, Austria: International Institute for Applied Systems Analysis.

Pettingill, J. (1998), 'Development opportunities for the Australian sawntimber and wood based panels industry', in ABARE (ed.), *Outlook 1998*, Proceedings of the National Agricultural and Resources Outlook Conference, Vol 1, Commodity Markets and Resource Management, Canberra: ABARE, pp. 308-12.

NON-WOOD BENEFITS

10. The Importance of Non-Timber Products in Tropical Small-Scale Forestry

S.R. Harrison and J.L. Herbohn

This chapter examines non-timber products arising from native and planted forests in the tropics, and associated issues including tenure, product marketing and rent sharing. The discussion draws on locations visited by the authors and their colleagues in India, the Philippines, China and Australia. It is apparent that non-wood benefits can be relatively large, and make a major contribution to welfare of low-income traditional communities. However, the utilization of these products needs to be considered in the light of sustainable harvest levels and equity issues. In developed countries, forest products tend to be less important than forest services.

10.1 INTRODUCTION

While plantations typically are established for the dominant use of producing timber, native forests are more frequently managed under a multiple use regime to provide a variety of products and services for local communities. In tropical areas, particularly in developing countries, communities often have a strong attachment to and dependence on forests.

Various nomenclatures are used to describe forest products other than timber. An early expression was *minor forest products* (MFP), although it became recognized that the importance of these products to traditional communities is considerably more important than implied by the term 'minor'. Other terms include non-timber (or non-wood) forest products (or forest benefits), hence the acronyms NTFP, NWFB and so on. Mallik and Panigrahi (1998) noted that minor forest products was accepted terminology prior to 1988, and now use the expression 'non-timber forest produce'. 'Non-timber' could be taken to include fuelwood (usually a highly important forest product), whereas non-wood probably would not. 'Benefits' is a broader concept than 'products' or 'produce', since it can encompass forest services. The focus of this chapter is on 'non-timber forest products'. Utilization of these products in India, the Philippines, southern China and northern Australia is examined. Some of the limitations on harvesting of NTFPs are reviewed.

10.2 TYPES OF FOREST BENEFITS IN DEVELOPED VERSUS DEVELOPING COUNTRIES

Tropical forests are by and large located in developing countries, those in North Queensland in Australia being an exception. The motives for forest management in developing countries can be quite different from that in more developed countries. The conservation of tropical forests tends to be easier in the latter, where it is more affordable and more enforceable. The indigenous population tends to be well protected by the social welfare net, and not to live in or be dependent for a livelihood on native forests, hence the extent of harvesting of NTFPs is typically low. Often, there is greater emphasis on other non-wood forest benefits (NWFBs), particularly environmental services and recreation. Indeed, much tree planting in developed countries takes place for environmental purposes, such as riparian revegetation and landscape aesthetics. The community have ceased to be hunters and gatherers, lack indigenous knowledge about usefulness of forest products, and have come to rely on a high level of services, expecting products to be available neatly presented on supermarket shelves.

In contrast, in developing countries and particularly in traditional (or tribal) areas, there is a strong association between the community and the forest, and heavy reliance for fuelwood, food, saleable products, medicines and so on. Forests play a major role in the subsistence and security of the community. Over time with deforestation and dispossession following colonisation, the welfare of traditional communities dwelling in and near forests has been adversely affected, and programmes are now being developed to improve access to the traditional common property resources of native forests, particularly in Asian countries. Various forms of community forestry have been devised to ensure the continued access to, and more sustainable management of, forest resources by local communities, e.g. see Harrison *et al.* (Ch. 18, this volume).

Various classifications have been developed for forest products and services other than timber production. The National Commission on Agriculture (1976) in India (reported by Mallik and Panigrahi (1998) classified NTFPs into 11 categories. Mallik and Panigrahi (1998) themselves adopted a five category system, viz. vegetables and oilseeds; food items; tan stuff and dye stuff; drugs; and miscellaneous. A modified classification is adopted here, presented in Table 10.1, which illustrates the range of products available to traditional communities.

Table 10.1 Some categories of non-timber forest products

Benefit category	Specific examples
Biomass fuel	Fuelwood, dung
Grazing, fodder collection	Feed for cattle and goats
Commercial crops	Rattan
Food	Mushrooms, edible tubers, fruit, honey, wildlife/game, ingredients for beverages
Tools and utensils	Wooden ploughs, paper plates, weapons
Medicinal products	Broad range of herbal cures
Other products	Food additives, kendu leaves, seeds, dyes
Living space, spiritual values, recreation	Forest dwelling and forest visits
Environmental benefits	Watershed protection, shelter, flora and fauna habitat

Non-Timber Products of Planted Forests

The availability of a wide variety of traditional forest products usually depends on forests containing a rich mixture native plant species, which is typical of native forests but uncommon in plantations. A variety of planted forest systems are adopted in tropical developing countries, and these vary in terms of the NTFPs they provide. Table 10.2 sets out some examples. A strong preference by smallholders is sometimes observed for planting fruit rather than timber trees. Many 'crop' trees grown in plantations also become important timber sources, including coconut and rubber trees. Agroforestry is widely practiced in developing countries, to gain complementarities between tree and crop production, e.g. combining timber or fodder trees with grain or vegetables. Similarly, rattan is best grown within a forest matrix. Environmental plantings on steep lands sometimes have an incidental food, fuelwood or timber harvest. Plantations of exotic timber species generally provide little in the way of NTFPs. This is one reason why introduction of species such as conifers and eucalypts in tropical developing countries, particularly as monocultures, is often resisted by local people.

Relative Importance of Non-Timber Forest Products

Non-timber forest products in aggregate make a major contribution to the livelihoods of poor people in developing countries in the tropics. According to Ames (1998, p. 107), 'CIFOR estimates that timber and other forest products provide 350 million people living in or around tropical forests with 50 per cent or more of their household needs and also directly provide 10 per cent of jobs in developing countries'. These products can be far more than supplements to the livelihood; in times of crisis, the store of food or saleable products in the forest may mean the difference between starvation and survival.

Table 10.2: Spectrum of timber versus non-timber forest benefits

Forest type	Forest uses
Fruit trees	Almost exclusively for produce, with some terminal fuelwood value
Rubber or coconut plantation	Mainly for produce during plantation economic life, but with possibility of undercropping and with timber and fuelwood yield at termination
Agroforestry, e.g. alley cropping	Trees not usually high quality timber, but used for shelter or nitrogen fixing and fuelwood
Timber species with rattan	Trees act as physical support to establish other important crop
Steepland and riparian forests	Designed for environmental services, with possibility of limited timber harvest
Timber plantation co-managed by forest service and community	NTFP may be highly important, but so is timber
Conifer or eucalypt monoculture	High timber yield but little non-timber value

While a large number of NTFPs have been recognized, a few products tend to dominate in the tropics and become subject for commercial development. de Jong and Utama (1998) report a process of identifying candidate non-timber species for development of a marketing and business plan in West Kailmantan. The products are ranked according to a set of ecological, socio-economic and enterprise criteria. Out of 14 target products, highest overall rankings were obtained for rubber, domesticated pigs then bamboo. According to de Beer and McDermott (1996), bamboo and rattan are widely regarded as the most important NTFPs in Asia.

While a complex timber production pipeline often exists in developed countries, an even more complex pipeline sometimes arises for NTFPs in developing countries. Belcher (1998, p. 57) noted that a forest product 'may change hands many times and go through a series of processors before it reaches the final consumer', referring to this as a production-to-consumption system (PCS). The PCS approach is used by the International Network for Bamboo and Rattan (INBAR) to assess the opportunities and constraints in marketing of NTFPs. Notably, there is often a high degree of vertical integration in the processing of NTFPs.

10.3 CASE STUDIES OF UTILIZATION OF NON-TIMBER FOREST PRODUCTS

Some comments are now made about the utilization of NTFPs in specific areas of four countries – viz. India, the Philippines, China and Australia – drawing where possible on fieldwork of the authors and co-workers.

Case Study 1: Non-timber Forest Products and their Commercialisation in Orissa and West Bengal, India

The benefits of non-timber forest products have been a research focus of the Nabakrushna Choudhury Centre for Development Studies in Bhubaneswar, Orissa[1]. Mallik and Panigrahi (1998, p. 101) noted that these 'have been able to meet the subsistence needs of food, fodder, fuel, household needs, agricultural implements and medicinal plants and herbs to a very great extent'. They also noted that 'the flora and fauna, streams and waterfalls, springs and mountains in the study region seem to have created a religious, emotional and recreational environment to the forest dwellers since time immemorial' (p. 101). They further commented on the remarkably wide knowledge and experience of basic usefulness of medicinal plants and also uses and environmental impact of the varied flora and fauna.

Mallik (undated) discussed the importance of the leaf of the Kendu tree (*Diospyros malanoxylon*) to tribal people in Orissa. This is a good example of a 'minor' forest product where the scale of operation has required government attention to marketing. Kendu leaves are used as wrapping for tobacco for preparation of beedi, because of the flavour, flexibility, feather texture and resistance to decay. Pruning and trenching of trees to ensure high quality leaves is undertaken by the Kenduleaf Division of the Forestry Department.

Kendu leaf accounts for 67% of the revenue of NTFPs from Orissa, and provides employment for thousands of forest dwellers and beedi makers, in the collection, drying, storage, processing and binding of leaves, and beedi rolling. 'Though seasonal, this has a great bearing on the household economy of innumerable poor forest dwellers and tribals in particular in the remote forest areas, whose principal source of livelihood is their sustenance from plucking of [kendu leaves]' (Mallik undated, p. 4).

In the past, the government operated leases, public auctions and tender calls, and a small number of merchants paid a token royalty to the Orissa government. The government entered into collection and processing of kendu leaf, with the *Orissa Kendu Leaves Control and Trade Act, 1961*,

---

[1] The centre has a collaborative research agreement with the Department of Economics of the University of Queensland.

which operated until 1972. The industry was then nationalised under the *Joint Scheme of Operation* in 1973, when 'the Forest Department was entrusted with the task of collection and the Orissa Forest Development Corporation was entrusted with the responsibility of disposal . . . by periodic auction sales' (Mallik undated, p. 4). The motivation for nationalisation was both to secure the government share of economic rent and to safeguard the welfare of the primary pluckers, who had been exploited by the private traders. The *Kendu Leaf Advisory Committee* fixes the purchase price annually. State revenue increased from Rs2.9 M in 1972 to Rs72.8 M in 1993. Other states which have large kendu leaf industries include Madhya Pradesh, Bihar and Maharashtra. The sales network for kendu leaf extends to Pakistan, Bangladesh and Sri Lanka.

In West Bengal also, forest products are important for community livelihood. A visit by one of the authors to Santal villages near Midnapore in West Bengal revealed the importance of the forests in everyday life. Villagers are involved in a joint forest management group. Leaves of the sal tree (*Shorea robusta*) are collected from the forest, and stitched together with twigs by women and children, to make plates of about 35cm diameter. About five leaves are used to make each plate. After drying, the plates are formed into stacks about 1m high, and transported by bicycle to a local merchant. The plates, which are disposable and biodegradable, are highly regarded and used at weddings and other special occasions. In one village, sale of sal plates was the major revenue source. Individual sal leaves are also used as containers for rice-based snacks for train passengers.

Cattle and goats are grazed in community forests and government eucalypt plantations. Collection of fuelwood is a prominent activity in tribal villages, with heavy bundles of sticks and single poles being carried by headloading, and bundles of sticks being carried by bicycle. Smokers use Kendu leaves as wrappings for tobacco. Mushrooms and a local kind of potato were also obtained from the forest. An highly prized drink is made from the flowers of the mahua tree.

A by-product of grazing cattle in forests is cattle dung, which is important as a fertilizer and fuel source. Travelling in India, dung cakes drying in the sun are a common sight. According to Panigrahi (1999, p. 37), dung cakes as fuel 'are very popular because of their availability within the house and throughout the year, easiness in preparation of cakes and their storage for the rainy season, and cheap source of slow and continuous heat'. He noted that about 45 M t of dung cakes are used for cooking fuel annually, accounting for about 12% of biomass energy in rural areas.

Case Study 2: Non-timber forest products in the Philippines

As in India, a wide variety of products are obtained from native forests in the Philippines. Of note is the combined enterprise of reforestation with production of non-wood products, including alley cropping and the growing of rattan. Common property forestry is managed by groups of Filipino smallholders under the community based forest management (CBFM) programme. A community forest at Alcoy on the island of Cebu, managed by the KMYLB Agro-Forestry Cooperative, was visited by the authors. One of the objectives is to grow 162ha of rattan, which will allow value adding through production of furniture and craft work. Attempts to establish this on cleared land were unsuccessful, and it was realised that rattan is best planted with timber, where it is assisted by the shelter and climbing framework provided by forest trees. Elsewhere in the Philippines, on Palawan Island, Ames (1998) estimated costs and revenues from harvesting of rattan and other products by local tribes holding Certificate of Ancestral Domain Claim (CADC) tenure. In this case, little value adding took place and the returns to the indigenous community were modest.

Case Study 3: Non-timber forest products in Yunnan Province, China

Ren and Tisdell (1996) examined community-based forestry in Yunnan Province, southern China, noting that almost all nationalities living in the forest areas have a rich knowledge of the non-timber products, such as edible mushrooms, wild fruit and vegetables, bamboo shoots, and spice and medicinal plants. Other timber products include raw materials such as tannin, lac, aromatic oil, resin, cork, gum, fibre, dyes, starch, forage and green manure. A team lead by Dr Ren observed 56 kinds of wild vegetables at local markets. Of note was the collection of food additives and dyes, such as the pulverised dried flowers of *Gmelina arborea* (a common timber species in SE Asia) to dye glutinous rice for the New Year Festival. Also, various handicrafts made use of forest products, including wood carving, rattan wares, furniture, tissue paper and dyed cloth.

Case Study 4: Harvesting of Forest Products in the Wet Tropics of Australia

In Australia, indigenous aboriginal communities made use of forests as a food source (such as bunya and macadamia nuts, marsupials and other game, wittchety grubs), for medicines, bark for housing, and timber for canoes. They had knowledge of a wide variety of edible bush foods and other products. They were able to apply leaching treatments to make foods which are toxic in their initial state suitable for eating, such as seeds of black bean tree and macrozamia palm. Traditional forest foods are still obtained in some tropical areas, such as Cape York, where turtle eggs, large birds (e.g. emu, brolga), dugong, crocodile, wild fruits and palm shoots are harvested. European settlers introduced their crops and livestock, and made little use of this resource or indigenous knowledge. In recent years, there has been increased interest in 'native foods' by the wider community, and recognition of traditional hunting rights of the indigenous people.

A somewhat different example of utilization of a non-timber forest product in a developed country relates to the harvesting of the feral pig (*Sus scrofa*), an introduced pest in tropical North Queensland. Feral pigs cause severe crop damage on some sugar cane and banana farms near rainforest areas, and environmental damage to forests and native wildlife the Wet Tropics World Heritage Area (WTWHA) in North Queensland. They have the potential to harbour diseases of domestic livestock (notably Foot-and-Mouth Disease), and be a vector for zoonoses (diseases transferable from animals to humans), including Trichononis and Encephalitis). These diseases could become a major concern in North Queensland, including in Aboriginal and Torres Strait Islander communities. Pigs also pose a physical risk to visitors to rainforest areas, and hence could deter ecotourism. There is an export market for 'wild boar' making it possible to treat feral pigs as a resource rather than simply a pest, although hunting is not permitted in the WTWHA.

It is notable that a number of reforestation programs have been instituted in tropical North Queensland, and one of the recognized benefits of these programmes is to provide habitat and corridors for native wildlife. It is now being recognized that reforestation, and plantings designed to protect riparian areas, can also lead to the negative externality of providing habitat for feral pests including wild pigs adjacent to crop land.

10.3 SOME ISSUES CONERNING THE ROLE OF NON-TIMBER FOREST PRODUCTS

While the utilization of non-timber forest products has been widely embraced as a welfare measure and alternative to unsustainable logging, there are some important limitations of this activity.

Relative Importance of Wood and Non-timber Products

Consideration of NTFPs implies a focus on management of forests to maximize the overall community benefit in the long run. The optimal forest use in terms of timber and non-wood products (and services) will depend on the technical relationship between these output groups, and relative prices. Ideally, the harvesting of non-timber products will not be highly competitive with timber production. Simple economic logic would suggest that there are some situations where timber production is a preferred forest use and others where a concentration on non-timber products brings greater societal benefits. More formally, these issues may be examined using the economic concepts of production possibility frontiers and isobenefit lines between timber and other forest products.

Examples of both complementarity and competitiveness can be readily identified. In that there is a tendency for inadequate thinning in farm and community forestry in developing countries (e.g. Harrison *et al.* 2000), collection of house poles and stakes may improve timber yield. Up to a particular grazing pressure, livestock and forestry are often complementary activities, the animals reducing the herb competition with trees. However, high stocking density of goats can lead to forest damage through herbivory and soil compaction by hooves. Collection of fruits and non-tree products such as mushrooms will be largely independent of timber production. Similarly, collection of dead timber will have little effect on tree growth. On the other hand, collection of green timber for fuelwood, and heavy harvesting of leaves (e.g. sal, kendu), could be expected to reduce timber yield, as could pruning and root treatment to produce multiple stems.

Often utilization of NTFPs will have a complementary relationship with timber production because of social rather than biological factors. When villagers are given greater property rights to forests, they will have greater incentive to manage the forests sustainability, with less illegal and premature logging.

Relative benefits will depend on both priced and unpriced values of forest products. As well, when income-distribution effects are considered, there may be a weighting towards non-wood products since these improve the livelihoods of large numbers of low-income beneficiaries.

Sustainability of Harvest Rates of NTFPs

Just as in the case of timber production, it is possible to harvest non-timber products at an unsustainable rate. As population increases and areas of native forest shrink, and when communities are under pressure in terms of revenue generation or food supplies, there is a tendency to increase the rate of harvest. In that small enterprises are often developed around the utilization of NTFPs, a related question is whether these enterprises are sustainable. Arnold and Riuz Pérez (1998) noted that many NTFPs encounter boom-bust cycles and argue that more attention should be given to understanding the limitations on NTFP conservation and development.

Further, they pointed to the need to analyse NTFP issues on a case by case basis, because of the high variation between particular cases.

As well as ecological sustainability, the economic sustainability of harvesting and marketing of NTFPs is sometimes called into question. Some of the issues are noted by Ames (1998). Forests are often characterised by poor roads and limited accessibility, leading to high transport costs for supplies, raw materials and finished products, and limited access to skilled labour markets. Products are often located in isolated areas far from the homes of collectors, so collectors may need to spend weeks in the forest, preventing them from holding other jobs, and may require cash advances from traders which can be difficult to pay off later. Products of high market value often have short and infrequent harvest seasons.

Social Equity Issues

A number of social issues can arise with respect to use of NTFPs:

Gender and age issues in utilization of non-timber forest products. Often the gathering of forest products and shepherding of livestock is performed by women and children. According to Kaur (1991, p. 2) in relation to India, '[t]he minor forest product economy . . . which is dominated by women, has never been the focus of government policy . . . Social forestry projects tend to be oriented to cash crops, which mostly benefit men'. Women are involved in small-scale forest-based industries, such as bidi-rolling and basket making, and also carry out some of the most arduous tasks. When fuelwood is scarce, women have to travel long distances and spend long hours in locating timber sources and headloading. Kaur (1991, p. 42) commented that 'women may have to walk up to 10 kms and spend five hours a day to gather firewood. Often the chore is peformed under dangerous conditions as when women carry 20 kg loads in the hills or when they gather firewood in both fear and defiance of the forest guard'. Sale of forest products is often a critical source of cash for women. Kaur (1991, p. 23) further notes that the subsistence economy is 'entirely dependent on the forests', and that shifting cultivation, which has been described as a 'female farming system' persists in tribal forest areas of India. Hence improving the access to forests and forest products can be critically important for equity reasons. As noted by Ames (1998, p. 108), 'selling forest products has become an important source of livelihood for those who are restricted by geography, tribal affiliation, capital requirements or gender roles, and therefore cannot participate in other cash-generating activities'.

Rent capture by middlemen. Mallik and Panigrahi (1998) noted problems in commercialisation of forest products in tribal areas of Orissa, India. When villagers sell these products to agents, they often receive what is only a small fraction of the final price. The government of Orissa has been concerned with the welfare of tribal people and exploitation by middlemen, and various measures have been trialled to improve the share of returns by the tribal people (as discussed earlier).

Dispossession through commercialisation. While the commercialisation of NTFPs can have great benefits for forest-dependent communities, it also carries risk. If products are sold outside the community, then those who have been traditional users within the community may not be able to acquire these products. That is, commercialisation can dispossess some of the traditional users, who operate outside a market economy. Sometimes it will be people from outside the community who will gain from commercialisation, particularly the traders or middlemen.

Formation of elite groups. Community ownership of forest resources can lead to formation or entrenchment of elite groups. If members of a particular group are granted common property access and development rights to a forest area, it would be expected that they would exclude non-members from access to the forest, including some who previously had traditional access. In this way, joint or community-based forest management can favour particular groups and increase inequalities in rural areas.

10.4 CONCLUDING COMMENTS

In tropical developing countries, some communities have traditionally relied heavily on non-timber products for their livelihoods, and community co-management with government of native forests has been developed as an institutional means of restoring property rights to these resources. A wide variety of non-timber forest products are utilized, as a source of food and other household products, for grazing and as security in case of emergencies. Forest products can provide an important source of cash income, and are sometimes commercialised, in which case marketing arrangements are needed to protect the welfare of primary collectors. Ulilization of non-timber forest products can play an important role in conserving native forests. However, constraints on utilization arise with respect to ecological and economic sustainability, and equity issues can arise. In developed countries, there is less emphasis on forest products and livelihood and more emphasis on environmental and recreational services.

REFERENCES

Ames, M. (1998), 'Assessing the profitability of forest-based enterprises', in E. Wollenberg and A. Ingles, (eds), *Incomes from the Forest: Methods for the Development and Conservation of Forest Products for Local Communities*, Bogor: CIFOR and IUCN, pp. 107–136.

Arnold, J.E.M and Riuz Pérez, M. (1998), 'The role of non-timber forest products in conservation and development', in E. Wollenberg and A. Ingles, (eds), *Incomes from the Forest: Methods for the Development and Conservation of Forest Products for Local Communities*, Bogor: CIFOR and IUCN, pp. 17–41.

Belcher, B.M. (1998), 'A production-to-consumption systems approach: Lessons from the bamboo and rattan sectors in Asia', in E. Wollenberg and A. Ingles, (eds), *Incomes from the Forest: Methods for the Development and Conservation of Forest Products for Local Communities*, Bogor: CIFOR and IUCN, pp. 57–84.

de Beer, J.H. and McDermott, M.J. (1996), The Economic Value of Non-timber Forest Products in Asia, 2nd ed., Amsterdam: Netherlands Committee for IUCN.

de Jong, W. and Utama, R. (1998), 'Turning ideas into action: planning for non-timber forest product development and conservation', in E. Wollenberg and A. Ingles, (eds), *Incomes from the Forest: Methods for the Development and Conservation of Forest Products for Local Communities*, Bogor: CIFOR and IUCN, pp. 43–55.

Harrison, S.R., Herbohn, J.L., Dart, P.J. and Brown, S.M. (2000), 'Sustainable small-scale forestry in the Philippines', in Harrison, S.R., Herbohn, J.L. and Herbohn, K.F. (eds), *Sustainable Small-scale Foresty: Social and Economic Analysis and Policy*, Cheltenham: Edward Elgar, pp. 165–178.

Kaur, R. (1991), *Women in Forestry in India*, Washington DC: Women in Development Division, The World Bank.

Mallik, R.M. (undated), *Procurement and Marketing of Kendu Leaves in Orissa: A Study of Deprivation and Benefits to Primary Collectors*, Bhubaneswar Nabakrushna Choudhury Centre for Development Studies.

Mallik, R.M. and Panigrahi, N. (1998), *Non-Timber Forest Produce Collection: Benefits and Management in Orissa*, Bhubaneswar, Orissa: Nabakrushna Choudhury Centre for Development Studies.

Panigrahi, M.R. (1999), 'Importance of bioenergy in sustainable forest management', *Ori-Forest*, Issue 5, June, pp. 36–41.

Ren, Z. and Tisdell, C. (1996), 'Developing community-based forestry in the uplands of Yunnan: dictates of environment and socio-economics', in C.A. Tisdell and J.C.H. Chai (eds), *China's Economic Growth and Transition: Macroeconomic, Regional, Environmental and Other Dimensions*, Economics Conference Monograph No 2, Brisbane: The University of Queensland, pp. 401–414.

11. The Valuation Impact of Reforestation on Degraded Farm Land

S.R. Harrison, N.H. Stehn, S. Boydell and J.L. Herbohn[1]

Rainforest trees planted on farmland can provide various aesthetic, environmental and habitat benefits, as well as having potential to produce valuable cabinet timber. Although there is a widely held community belief that the planting of native rainforest trees on degraded farm land will increase the value of that land, and this is one of the arguments advanced to promote farm forestry, rural valuers and real estate agents dispute this view. Survey evidence indicates that even well-established trees may have only a small impact on land value. This appears to be due, at least in part, to lack of information as to potential returns from farm forestry.

11.1 INTRODUCTION

There appears to be a widely accepted belief that trees increase the asset value of land, provide enhanced security against which to borrow funds, give rise to aesthetic values, provide habitat for native birds and animals, and in general yield considerable private benefits to landholders. For example, Greening Australia (1996, p. 32) refers to 'property value enhancement'. Presence of trees is often used as a feature in advertisements for land for sale. Loss of rural trees due to Eucalypt dieback has been found to impose considerable cost on landholders (Harrison and Sutherst 1997).

In contrast to these views, there is anecdotal evidence that presence of trees may reduce land values. Landholders sometimes take the view that the

---

[1] This chapter summarises research conducted by Ms N.H. Stehn as part of a M.Agr.Econ.Studs. thesis and in conjunction with the other co-authors. Financial support of the Rainforest CRC and North Queensland Joint Board is gratefully acknowledged.

cost of clearing should be deducted in arriving at land value, and that tree planting increases habitat for feral animals and is a source of weed seed for cropland. Tree cover has nuisance value where a landowner wishes to build or subdivide. Similarly, trees may not be the 'highest and best use' of land from a valuation viewpoint. While reforestation has generally been considered an environmentally friendly activity, concern was expressed for the visual disamenity of plots of exotic conifers on highly attractive natural mountain landscape in Wales by Price (1989). Similarly, Drake (1992) found a willingness-to-pay to prevent loss of agricultural land to spruce forest in Sweden of about 78 ECU (approx. $A100) per person per year.

There is a general lack of agreement on whether productive and environmental improvements to land are fully reflected in land values. Economic studies have revealed that the land market is generally insensitive to varying levels of soil erosion, probably due to imperfect information on extent and effects of erosion (Ervin and Mill, 1985). Gardiner and Barrows (1985) found that soil conservation investments are not capitalised into farmland prices, which they also attributed to imperfect information available to buyers. On the other hand, King and Sinden (1988) found no evidence that the market undervalues land conservation measures relative to the expected yield effects of those measures.

This chapter reports findings of a study of rural valuers and real estate agents on impact of reforestation on land values in North Queensland, Australia. Further details are available in Stehn (1997).

11.2 VALUATION METHODS FOR FARM FORESTRY

Literature on the valuation of forested lands indicates that the major methodologies are the market comparison approach, the summation method (or cost compounded method) and the income capitalisation method (e.g. Eves, 1996a,b; Laing, 1993; Lewis and Ferguson, 1993). The method of valuation chosen will depend on the whether the trees have any commercial value, and if so, the proposed harvest regime.

The market comparison approach, whereby a valuation is determined by reference to the market-value of similar properties, faces a number of difficulties with respect to data availability and quality. There is diversity in the forest land's physical features (such as soil condition and topology), locational features (such as climate and distance to sawmills) and in the management practices (such as species mixture, stand density and pruning and thinning regime). This diversity affects the forest's growth rate and the quality of the timber. Sales on the open market of large tracts of forest are rare, due to ownership being largely confined to large corporations and government. Also, sales of large areas of forest may be conducted privately rather than through the real estate industry. When properties with small forest areas are sold, details of forest value usually are not mentioned specifically in valuation records. Market comparison techniques require consideration of a large number of factors to ensure the subject and reference forests are comparable. To achieve this, Laing (1993, p. 498) recommends use of forest description models listing details of comparable forests which may affect their value. A detailed list of factors which should be considered in the market comparison approach is provided by MacClaren (1993) (reported by Eves, 1996b, pp. 5–6). The complexity of the information required and the likely need for substantial adjustments for variations between the comparable properties and the subject property may reduce the weight which a court gives to valuation evidence based on this technique.

The *summation or cost compounded method* assumes that the value of a property is the sum of the values of its constituent parts. Components, such as buildings, are valued at their reproduction cost less adjusted for depreciation. The value of the land component is usually determined by the market comparison approach. A variation of the summation or cost compounded method has been used to value orchards prior to production, and Eves (1996b, p. 4) considered it may be used for valuing pine plantation forests not more than eight years old. The method involves compounding costs of plantation (or orchard) development to date, such as planting and tending costs, and valuing the land separately. Under this methodology, the compounded cost of establishment and tending the trees may exceed any value which the market may attribute to the trees. Hence, the value may need adjusting to reflect market experience (MacClaren, 1993, reported by Eves, 1996b). An extension of this approach is to compound historical costs to present values and add the private non-timber values such as amenity value.

The *income capitalisation method* capitalises the present value of the future income stream associated with the property and uses this as the basis for valuation. Hence, an *income capitalisation* rate may be applied to the annual income level. The rate by which future incomes are discounted reflects the investor's rate of time preference of the present over the future, the opportunity cost of the investment and the investment's degree of risk. Forests which are selectively logged (such as some native hardwood forests) may be valued using this methodology (Rost and Collins, 1993, pp. 373–377). In the case of large public forests, the annual income stream may be reasonably regular if a constant level of harvesting is conducted annually across the entire forest. More often, however, income is irregular. While predictions could be made based on the income from any one year, or group of years, this may not be representative of income from other years. Rost and Collins (1993) suggest that for valuing land with other than minor areas of marketable timber, a forest survey should be conducted by a qualified forester to determine the quantity and value of the timber. For small areas of land, the variety and quantity of each useable tree in the area should be determined, with deductions made for defects. For larger areas, a sampling technique or 'timber cruise' may be used in which some only of the trees are examined for timber quality and volume.

The value of a timber stand may be estimated in terms of *net present value* and *land expectation value*, particularly where cash flows are reasonably predictable, e.g. for exotic conifer plantations (c.f. plantations of non-traditional rainforest species). Acceptance of discounted cash flow (DCF) analysis by the real estate profession or the courts has been gradual. The DCF method was rejected by Jacobs J. in the High Court Decision of *Albany & Ors.* v. *The Commonwealth of Australia* (1976) 25 The Valuer, No. 7. In September 1996, the Australian Institute of Valuers and Land Economists issued a practice standard for use of DCF analysis for income producing property, development projects and land subdivisions (Boydell and Gronow 1997, p. 4). Eves (1996a,b) used DCF analysis in an appraisal for vendors of a New South Wales softwood pine plantation, achieving a property value estimate within 5% of that obtained when the property was sold three months later at a public auction where there were three genuine bidders. He concluded that DCF analysis may be appropriate as a primary method of valuing hardwood and softwood plantations, to be checked against available market data, if any. Income methodologies may be appropriate for valuing plantation forests which will be clearfelled at the end of the rotation. Laing (1993) recommends, in the case of New Zealand's exotic pine plantations, that income methodologies be used for stands that are at least 11 or 12 years old.

The timber industry uses a variety of techniques to value its assets (Herbohn *et al.*, 1998). Often, these involve accounting methods and are designed to provide book values for forestry assets in financial reports. Such values may differ significantly from the forest's market value (Stanford 1993, p. 24). These techniques may not work well for small-scale mixed-species plantings of native species on private farm land.

11.3 RESEARCH METHODOLOGY

A survey was conducted to determine: (1) the effect if any, of reforestation with plantation rainforest cabinet timbers on land values in the Queensland Wet Tropics, and (2) the valuation methodologies employed by valuers and real estate agents to value reforested land. It was hypothesised that reforestation on degraded farm land would increase land value, and that the increase would be sensitive to a number of factors, such as stand age, species planted, distance from a timber mill and amount of protected native vegetation in the area. Also, it was anticipated that values would reflect both potential timber yields and private non-wood benefits such as soil protection and aesthetic and habitat values.

Initial investigation revealed that few land valuers or real estate agents have experience or confidence in valuing forestry, and drawing a random sample of experts was not feasible. Further, interviewing landholders could reveal perceptions and yield anecdotal data but these would not be closely linked to market realities. Finally, a group of only six experts was chosen, including two land valuers, three real estate agents and one landholder. Two additional landholders and a forester participated in recorded interviews only. Discussions were also held with other residents and landholders. While respondents had experience in land valuation or in the real estate industry in their particular locality (either on the Atherton Tableland or Cairns and the Daintree area), they had little experience in valuing reforested land. Most of the established plantings of native species (including hoop pine, red cedar and Queensland maple) are on government land and have not been the subject of private real estate sales. It is likely that some of the views expressed may have been based on a comparison with values for land with remnant rainforest.

The survey was conducted at Cairns and on the Atherton and Evelyn Tablelands in early December 1996. Respondents were posted a questionnaire for consideration before interview. This comprised structured questions, largely of a quantitative nature, and unstructured parts allowing respondents to provide comments on or expand on their earlier answers to the structured questions. The structured questions included presenting the experts with four forestry scenarios, and asking them to value both bare and forested land. Colour photographs were taken of farm forestry on the Atherton Tableland and of a residential subdivision in the heavily forested Daintree area. These photographs were scanned, and printed on a colour printer as part of the questionnaires, so as to help respondents to visualise the reforestation scenarios. The unstructured component of the questionnaire allowed respondents to make written comments in addition to answering specific questions. As well, oral comments were recorded with a microcassette recorder, and typewritten transcripts were made and analysed qualitatively. This approach allowed more detailed comments to be obtained and specific issues to be explored.

11.4 FINDINGS FROM THE SURVEY OF FOREST VALUATION EXPERTS

Effect of Reforestation on Land Values

Question 1 asked whether, in general, reforestation with cabinet timber species has an effect on the value of land, for trees of various ages. As reported in Table 11.1, most experts considered there is a positive relationship between plantation age and land value, though there may be little or no impact on land value for the first few years after planting.

Table 11.1 Effect of reforestation on land values (n=5)

Age of planting	Impact on land value			
	Decrease	No change	Small increase	Big increase
Young trees (to 5 years old)	0	3	2	0
Older trees (6 to 20 years)	0	1	2[a]	1[a]
Older trees (> 20 years)	0	1	2	2

Notes: a. One respondent answered 'between small and big increase'.

One respondent noted that a recent valuation report of rural land containing valuable timber actually gave separate (though low) consideration to the value of timber. A second respondent judged that reforestation would have no effect on land values. This person had considerable experience in valuing small blocks in the Daintree area where land prices appear similar for forested and cleared land. Another respondent said that regardless of tree age, the value of prime arable land may decrease if reforested. For trees up to 20 years of age, the respondent considered a small value increase may occur for reforestation of better quality grazing land if the trees are placed at an appropriate spacing to enable pasture to grow between them, i.e. an agroforestry system. For trees of all ages, this respondent judged that a big increase in value may occur for degraded land.

It may be that the existence of trees, even those more likely to have timber value, is not considered by buyers as enhancing land value, other than where the trees give the land aesthetic value. No respondents spoke of purchasers buying reforested land for commercial reasons. Two respondents referred to the 'superannuation' or 'rainy day' value of stands of timber on farming land. One of these indicated that in some cases farmers with valuable stands harvested them to pay for special expenses, such as their children's university education.

Effect of Changed Incentives and Community Awareness on Land Values

Despite efforts to re-establish a cabinet timber industry in North Queensland following the World Heritage listing, serious impediments to achieving a viable industry exist. Respondents expressed various views as to why this is so and suggested solutions. Question 2 asked 'Do you think that any of the following changes could increase the value of reforested land, and if so why?', followed by a series of reforestation incentive measures. Responses are summarised in Table 11.2.

Table 11.2 Effect of changed forestry incentives and community awareness on land values (n = 6)

Change	An effect on land value	No effect on land value
Clearer market incentives	6	0
Improved government incentives	5	1
Greater community awareness of economic benefits	3	2
Greater community awareness of environmental benefits	4	1
Greater community awareness of amenity value	4	1

Notes: a. One respondent was undecided for Questions 3–5.

All respondents thought clearer market incentives would increase the value of reforested land. Concerns were expressed at a lack of information on the economic viability of cabinet timber plantations. Also, there were concerns that local sawmillers are not paying sufficiently high timber prices to landholders and that prices would be further eroded by competition from lower priced imported timber. Most respondents thought further government incentives to encourage reforestation would increase the value of reforested land. One argued that government incentives are necessary most growers would not reap the benefits during their lifetimes. Forward selling contracts and joint ventures with government were considered desirable by this respondent. Another respondent thought emphasis should be given to development of a veneer timber industry for the high quality North Queensland timbers, with supporting government research and development. The potential of reforestation for future medicinal purposes and bush tucker was also mentioned.

There was general consensus that education is needed to make the community aware of the advantages to land of reforestation and of the potential economic returns. Some respondents expressed concerns about the lack of economies of scale of small timber lots. The view was also expressed that smaller timbered blocks are purchased for their aesthetic value rather than their future earning potential. One respondent thought that improved tax incentives would lead to increased plantings and hence increased impact on land values. The majority of respondents considered greater community awareness – including awareness by banks of the economic benefits of reforestation – would increase values of reforested land. Other factors which respondents thought may affect land values include the existence of research findings showing favourable profit margins for reforestation when compared with commodity crops, improved tax incentives for forestry operations and more certain harvest rights.

Factors Which may Affect the Value of Reforested Land

Question 3 probed views on the effect on value of reforested land, of a number of factors, as indicated in Table 11.3. Most respondents thought that size of the property would have some effect. One commented that 100ha would be needed for a viable forestry operation while another considered that a minimum of 50ha would be required. All respondents thought land quality would have at least a slight effect on the value of reforested lands, with two-thirds considering it would have a substantial effect.

Table 11.3 Effect of various factors on value of land (n=6)

Factor	Effect on value		
	None	Slight	Substantial
Size of property (ha)	1	1.5[a]	3.5[a]
Area planted on the property (ha)	0	2	4
Land quality	0	1.5[a]	4.5[a]
Closeness to an urban area	5	0	1
Closeness to a timber mill	1	3.5[a]	1.5[a]
Land zoning	2	1	3
Native versus exotic species planted	0	1	5
Monoculture versus mixed species planted	0	2	4
Area of native forest in the district	3	2.5[a]	0.5[a]

Notes: a. Responses of 'between slight and substantial' have been divided between the two categories.

Closeness to an urban area was considered to have no effect by all except one respondent. A concern was expressed that urban residents may protest in the future against harvesting. Views varied on whether closeness to a timber mill (hence lower transport costs) would affect forested land values.

Views also varied about the effect of land zoning, though it seems provided forestry is a permitted land use, it would have no effect. Most respondents thought that the growing of native species, rather than exotics, would have a substantial effect on land value. Two-thirds thought that mixed species plantings rather than monocultures would have a substantial effect. Most considered that the area of protected remnant native forest in the district would have no effect or only a slight effect, on land values.

Ease of Sale of Reforested Degraded Land

The experts were asked whether they thought degraded land in their region becomes easier to sell when rainforest trees are planted on it, for plantings of various ages. As indicated in Table 11.4, most believed that land would be easier to sell, provided trees are well established (6 years or older).

Table 11.4 Ease of sale of degraded land if reforested (n=5)

Saleability	2 year old trees	6 year old trees	15 year old trees
Land easier to sell	2	5	5
No change in salability	3	0	0
Land more difficult to sell	0	0	0

Effectiveness of Native Trees as a Selling Feature of Land

A ‘yes’ or ‘no’ answer, plus comments, were sought on the following question: ‘Is the presence of native trees used as a selling feature in advertisements for the sale of land in your area?’. Follow-up questions concerned speed of sale and price effect. All respondents considered the presence of trees is sometimes used as a selling feature in advertisements for the sale of land in their area. One respondent thought that the presence of trees probably would not be used for promoting the sale of primarily agricultural production.

There was unanimous agreement that presence of native trees would assist in speeding up sales and in obtaining higher prices, although provisos were made concerning age of trees and potential returns from timber. The use of native trees as a selling feature indicates that buyers of some types of properties attribute some value to the presence of native trees. It is likely that real estate agents adjust their advertising to suit prospective purchasers.

Methodology for Valuing Reforested Land

The most vexed issue for professionals valuing private land reforested with cabinet timbers is the choice of valuation methodology. Question 6 stated 'If you were valuing an area of reforested land, how would you estimate the effect of trees on land value – for land with plantings of two years, six years and 15 years of age?' While all respondents were prepared to discuss the advantages and disadvantages of various methodologies, only three gave definite indications of the methods they would use. No respondents considered that they would estimate the effect of 2-year-old trees on land values. For land with 6-year-old trees, one respondent suggested that establishment plus maintenance costs should be used but it should be adjusted downwards, rather than compounded, to better reflect market value which is usually lower than the sum of costs incurred to the date of sale. Another respondent suggested costs plus amenity and other values, with the resulting estimate tested against results achieved by market comparison. A third respondent suggested use of costs plus amenity values.

For land with 15-year-old trees, only two respondents were prepared to suggest a methodology. One supported use of discounted cash flows whereas the other suggested market comparison. The former respondent held some reservations about the accuracy of DCF-based valuation because of uncertainty about yield and harvest age. A third respondent said a forestry expert would be employed to value forested land if trees were ready for harvest, but if the trees were only three or four years of age, no additional value would be added to the land apart from possibly amenity value. One respondent was waiting for researchers from the Rainforest CRC to develop a financial model for the timber valuation problem faced by land valuers.

Change in Value of Reforested Land

Question 7 asked 'Do you think the relationship between reforestation and land values has changed over the last 10 years?' Apart from one respondent with experience in the Daintree area, all agreed that trees are now more likely to increase land value. One respondent considered public awareness of the benefits of reforestation is increasing, but the public is not fully aware of the long-term financial and environmental benefits. Another said landholders know the value of remnant vegetation because some landholders have harvested timber prior to property sales. One respondent considered that while reforestation increases land values, landowners will still not get a high price for good rainforest blocks. Another said trees are increasingly being seen as a superannuation investment and that in other cases people grow them for 'feel good' (about the environment) reasons.

11.5 EVALUATION OF REFORESTATION SCENARIOS

The second part of the questionnaire involved four forestry scenarios. The first three sought to determine change in land value arising from reforestation on the Atherton Tableland in three circumstances, viz. reforestation with commercial timber species, reforestation with commercial timbers in a district where there is little remaining rainforest, and reforestation with non-commercial timbers on a degraded riparian area. The final scenario sought values for small rainforest blocks in the Daintree area. Respondents were first asked to value the bare land component of the property in terms of their best estimate, lowest value and highest value, then make the three value estimates with trees present.

Responses to Scenario 1. The first scenario concerned a 150ha dairy farm located near Atherton on fertile basaltic soils. The land is relatively level, with ample water but contains a 5ha area of degraded unproductive land of moderate slope. The district has a large area of protected native rainforest. The farm has all the necessary equipment for dairying operations and a milk quota. Valuations on the land component are summarised in Table 11.5(a), while values for the land carrying six-year-old trees are reported in Table 11.5(b). The mean of best estimates of bare-land value is $4,750/ha and the range is $3,750/ha to $5,375/ha. Three respondents thought there would be no change in land value with 6-year-old trees, one thought the increase would be small and one thought there would be a larger increase in value (of $1,000/ha). The mean best estimate of change is an increase of $270/ha. No respondents considered that the land would decrease in value.

Responses to Scenario 2. Scenario 2 sought respondents' views on whether there would be any change in land values nominated in answer to Scenario 1 if only a small area of rainforest remained in the district (c.f. the large area in the Wet Tropics of Queensland World Heritage Area). All respondents considered that a lack of rainforest in the district would not change the valuations suggested under Scenario 1. One respondent replied that it makes no difference to dairy farmers what rainforest is in the district.

Responses to Scenario 3. Scenario 3 presented the same facts as Scenario 1 except that the 5ha of degraded unproductive land is along a creek, where the land is suffering erosion and bank slumping. Estimated values for the bare land component varied considerably, ranging from $500/ha to $4,500/ha. One respondent valued the riparian area more highly per hectare than the property generally, whereas other respondents valued it considerably lower. If this land were reforested, respondents considered values would increase for that area. Best estimates of the increase ranged from $200/ha to $500/ha. No respondents considered that riparian land values would decrease as a result of reforestation.

Table 11.5(a) Value of land component without reforestation under Scenario 1 ($/ha) (n=4)

Best estimate of land value	Lowest value likely	Highest value likely
Mean = $4,759	Mean = $3,750	Mean = $5,375
Range: $4,000 - $6,000	Range: $2,500 - $5,000	Range: $4,500 - $7,000

Table 11.5(b) Value of land component with six year planting under Scenario 1 ($/ha) (n=5)

Type of change (increase or decrease)	Best estimate of change in value	Smallest increase (largest decrease) likely	Largest increase (or smallest decrease) likely
No change – 3	Mean = $270 increase	Mean = $140 increase	Mean = $300 increase
An increase – 2	Range = increase of $0 - $1,000	Range = increase of $0 - $500	Range = increase of $0 - $1,000

Responses to Scenario 4. The last scenario concerned a 3ha residential block in the Daintree area, located within rainforest. The block is connected to electricity and water supplies. Respondents were asked their valuation of such a block if it had a small house pad cleared in the rainforest, with access via a discrete low-impact access track off the main gravel access road. Respondents were also asked for their valuation if 0.5ha has been cleared and there is no rainforest barrier between the house and the main access road. Best estimates of values for the block with the smaller cleared area range from $28,500 to $70,000. For the more cleared block, values range from $25,500 to $55,000. Two respondents with more experience in Daintree land sales estimated no change in value between the well forested and the more cleared block.

11.6 DISCUSSION

Few people have confidence in placing values on farm woodlots, and hence it is difficult to obtain reliable information of the impact of trees on land values. Further, there does not appear to be any consensus as to the appropriate valuation methodology, approaches suggested including estimation of establishment cost plus non-wood values, discounted stumpage value and market comparison.

Evidence from the survey appears to be somewhat negative relative to the prior hypotheses. Clearly, young trees add little or no value to land. Even for trees aged 20 years or more, the increase in value may not be great. Effect of presence of trees on land value depends on land use. For profitable commercial farms, trees beyond those needed for shade and shelter purposes appear to add little value. Similarly, for residential blocks buyers appear to favour open space. Trees may add greatest value for marginal farms on degraded land and for non-commercial and lifestyle holdings. Ability of trees to add value appears independent of the extent of protected public forest in the area, or distance from a timber mill. The evidence suggests that trees on farms are valued for non-timber benefits, and are used to good effect as a selling feature. The greatest effect on value is achieved by relatively large blocks of mixed native species.

The scenario analysis indicated that the increase in value for six-year-old trees relative to bare and degraded land is much less than the cost of plantation establishment (>$2000/ha). The land market does not fully reflect the benefits of tree planting. It would appear that lack of information about expected profitability of native timbers is an obstacle to these future revenues being factored into property valuations. In that a valuation could be challenged in a court of law, a reputable and consistent basis of valuing timber is required. Even greater difficulties can arise in placing values on non-market benefits such as aesthetic benefits and wildlife habitat.

In summary, contrary to popular opinion, reforestation may contribute little to enhancing land value. Hence, promotion of farm forestry on the basis of increasing land values, or the value of assets against which to borrow, may be misleading advice. To the extent that this is due to lack of information about future timber returns, a greater value premium may arise when financial models of rainforest timber species are further developed and more widely accepted.

REFERENCES

Boydell, S. and Gronow, S. (1997), 'The Australian Institute of Valuers and Land Economists (Inc) discounted cash flow practice standard', *Journal of Property Valuation and Investment*, **15**(1), pp. 58–69.

Drake, L. (1992), 'The non-market value of the Swedish agricultural landscape', *European Review of Agricultural Economics*, **19**(3), pp. 351–364.

Ervin, D.E. and Mill, J.W. (1985), 'Agricultural land markets and soil erosion: policy relevance and conceptual issues', *American Journal of Agricultural Economics*, **67**(5), pp. 938–942.

Eves, A.C. (1996a), 'Do valuers need to change', in *Meeting Change with Change*, published papers of the AIVLE Annual Rural Conference, Kooralbyn, pp. 19–21

July.

Eves, A.C. (1996b), *The Feasibility and Valuation of Plantation Timber: A Case Study of Discounted Cashflow in Rural Valuation*, Sydney: University of Western Sydney.

Gardiner, K. and Barrows, R. (1985), 'The impact of soil conservation investments on land prices', *American Journal of Agricultural Economics*, **67**(5), pp. 943–947.

Greening Australia (1996), *Farm Forestry in Australia: Integrating Commercial and Conservation Benefits*, Canberra: Greening Australia.

Harrison, S. and Sutherst, R. (1997), *Cost of Accelerated Eucalypt Dieback due to Climate-Change on the New England Tableland*, Cooperative Research Centre for Tropical Pest Management, Brisbane: The University of Queensland.

Herbohn, K.F., Herbohn, J.L. and Petersen, R. (1998), 'Accounting for forest assets: current practice and future directions', *Australian Accounting Review*, **8**(1), pp. 54–6.

King, D.A. and Sinden, J.A. (1988), 'The influence of soil erosion on farm land values', *Land Economics*, **64**(3), pp. 242–255.

Laing, A.P. (1993), 'Forestry valuation methods: a New Zealand perspective', *The Valuer and Land Economist*, August, pp. 498–503.

Lewis, N.B. and Ferguson, I.S. (1993), *Management of Radiata Pine*, Sydney: Inkata Press.

Price, C. (1989), Public lecture on visual impacts of forestry, Agricultural Economics Association Conference, Aberystwyth.

Rost, R.O. and Collins, H.G. (1993), *Land Valuation and Compensation in Australia* 3rd. edn., Marrickville: Australian Institute of Valuers and Land Economists Inc.

Stehn, N. (1997), *The Impact of Revegetation on Land Values*, unpublished M.Agr.Econ.Studs. thesis, Brisbane: The University of Queensland.

Stanford, I.D. (1993), *Softwood Plantations – a Valuation Insight*, unpublished undergraduate applied research project, Sydney: University of Western Sydney.

12. Economic Evaluation of Riparian Revegetation Options in North Queensland

M.E. Qureshi and S.R. Harrison

Riparian zones are instrumental in protecting water quality from nutrient runoff and preventing bank erosion and sedimentation. Riparian vegetation reduces off-site damage costs and provides flora and fauna habitat. A discounted cash flow analysis has been carried out as a part of multicriteria analysis of riparian revegetation options for Scheu Creek catchment in North Queensland, Australia. Various cost and benefit categories are identified, including costs of planting and establishing riparian buffers, and the benefit of reduced rodent damage and control expenditure, reduced flood damage to crops and reduced off-site damage. Riparian revegetation imposes major costs on landholders, and much of the benefits are gained by other stakeholders. A knowledge of the costs and benefits involved in riparian revegetation will assist decision-making regarding choice of revegetation policy and implementation measures.

12.1 INTRODUCTION

One of the major elements of integrated catchment management (ICM) in Queensland is riparian revegetation, i.e. planting trees and shrubs along streambanks. This promotes bank stability and channel integrity, habitat quality and filtering of sediments and their attached nutrients (Raine and Gardiner, 1995). Riparian zones also serve ecological functions such as providing wildlife corridors, organic carbon, shade and habitat structure (Herron, 1996), and contribute to protecting groundwater quality, as well as reducing off-site damage costs and providing flora and fauna habitat.

The research reported here forms part of a *multicriteria analysis* (MCA) of revegetation options for the Scheu Creek sub-catchment of the Johnstone River Catchment in North Queensland. Scheu Creek runs for 13km through sugarcane land. Landholders along the creek have cleared trees off the banks and in some sections straightened the creek, increasing water velocity and risk of downstream flooding. Increased light intensity from the removal of riparian trees has provided ideal conditions for Para Grass (*Brachiaria mutica*) to flourish. Prolific growth of Para Grass has reduced creek depth and flow velocity. Soil erosion, bank slumping and siltation are common problems in the catchment, and the creek is heavily degraded due to sedimentation. Water quality has deteriorated and fish numbers are low; popular local swimming holes have become unusable due to sedimentation.

12.2 REVEGETATION POLICY OPTIONS

Differences of opinion arise as to the most appropriate form of riparian revegetation in the catchment. Some cane farmers prefer to retain Para Grass than to plant trees, because grass simply lays over in times of high river flow, protecting the stream bed and banks. Further, unlike trees it does not interfere with the turning of the cane harvesters, and does not compete with cane for plant nutrients. Bare areas of soil under trees are prone to erosion. Farmers are able to keep the water channel open and prevent siltation by periodic strip spraying with a herbicide. On the other hand, some canegrowers recognise that growth of the Para Grass impairs riparian habitat and aesthetics and causes a rat problem in their crops. Only large trees will bind banks and reduce summer water temperatures.

Bunn and Davies (1996, p. 39) examined the consequences of removal of riparian trees in Bamboo Creek (another tributary of the Johnstone River), noting that 'for a catchment of this size, the active channel width was estimated to be 10–15m; however, the stream was restricted by a matrix of Para Grass and sediment to a mean width of 1–2m'. They estimated that 20,000 tonnes/km of sediment was stored in Bamboo Creek, reducing the channel capacity from three times that needed to carry a one in 50 year flood to only one third of the capacity needed for a 1:50 flood event. They further noted that high discharge events inevitably lead to scouring of the channel and transport downstream of sediment and associated biomass.

With accumulation of sediment and decomposing plant material, the ability of Scheu Creek to carry flood flows is reduced. There is a danger that flood flows will spread over adjacent cane land, interfere with farm operations, reduce crop yield and perhaps cause soil loss. As well, there is removal of oxygen from the creek bed and a loss of creek biota.

Once established, riparian revegetation in the form of trees and shrubs, would shade the channel, reduce light and temperature and restrict or eliminate growth of Para Grass. This may mobilise sediment currently stored in the active channel, which could pose problems for coastal marine systems (Bunn and Davies, 1996). However, in the long run, riparian revegetation would lead to increased capacity of water flow and reduced severity of local (on-farm) flooding.

A number of revegetation options could be adopted for Scheu Creek, in terms of vegetation to be planted (e.g. trees or deep-rooted grasses) and width of planting strip. Depending on the location on the bank profile, various species have been suggested (Goosam and Tucker, 1995; Bell, 1996). Width of tree planting could be varied depending on the location along the creek and position in terms of straight reaches and meanders. After discussions with staff of the Department of Natural Resources (DNR) and the Johnstone River Catchment Management Association Inc. (JRCMA), four riparian revegetation options have been identified for Scheu Creek:

Option A: Three metre vegetated buffer on inside meanders, 6 metre buffer on outside meanders, 3 metre buffer along straight reaches.

Option B: Five metre vegetated buffer on inside meanders, 10 metre buffer on outside meanders, 5 metre buffer along straight reaches

Option C: Five metre vegetated buffer on inside meanders, 5 metre buffer on outside meanders, 5 metre buffer along straight reaches

Option D: Ten metre vegetated buffer on inside meanders, 10 metre buffer on outside meanders, 10 metre buffer along straight reaches.

There is currently a buffer strip outside the high bank of the creek, wide enough for vehicle traffic (about 4m). A cross section of the creek with proposed buffers under Option A is presented in Figure 12.1.

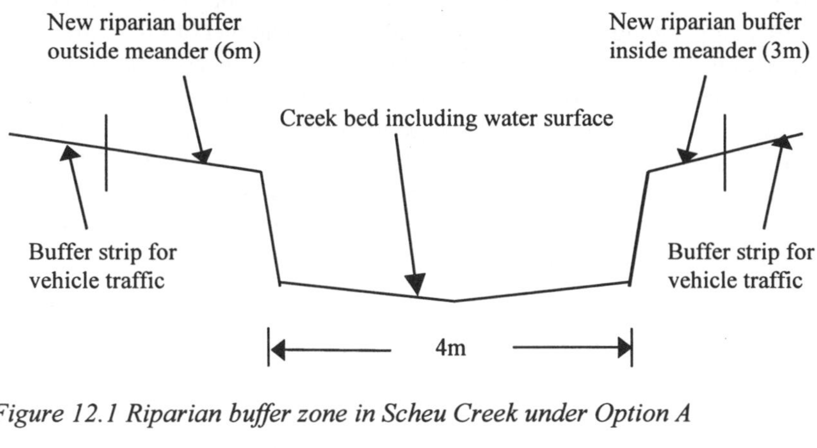

Figure 12.1 Riparian buffer zone in Scheu Creek under Option A

A geographic information system (GIS) has been used to estimate the total area which will be taken out of sugar production by riparian revegetation. GIS coverages were obtained from the Department of Natural Resources (DNR) in Mareeba. GIS software packages ArcInfo and ArcView have been used to identify meanders, estimate riparian buffer areas and produce maps. Areas of cropland lost under the four revegetation options are reported in Table 12.1. About 0.5ha of the creek is already planted to trees, and net land loss is reduced accordingly.

Table 12.1: Estimated loss of cropping land for riparian buffers in the Scheu Creek catchment

Revegetation option	Buffer area under option:			
	A	B	C	D
Total buffer area (ha)	7.2	12.3	10.2	20.5
Area planted (ha)	6.7	11.8	9.7	20

12.3 METHOD OF ANALYSIS OF RIPARIAN REVEGETATION OPTIONS

Riparian revegetation incurs costs and generates benefits over a number of years. Discounted cash flow analysis (DCF) provides an appropriate framework for evaluating economic impacts. Discounting reflects time preference for money, taking into account the fact that tree planting costs are invested immediately, while most of the benefits do not commence until trees have grown for several years. Revegetation options are evaluated over a 30-year planning horizon, using a real interest rate (net of inflation) of 7%. Economic analysis is performed from both a private perspective (that of cane farmers) and a social perspective (net cost for the whole community). Data on costs and benefits of revegetation policy options are drawn from Russell *et al.* (1990), Bell (1996), NQJB (1997), Canegrowers (1997), and subjective estimation by the authors. The analysis isolates those annual benefits and costs which change as a result of revegetation policy options relative to retention of Para Grass (i.e. *incremental cash flows*).

12.4 ESTIMATION OF THE COSTS OF RIPARIAN REVEGETATION

In establishing new riparian revegetation, costs involved include the opportunity cost of land no longer available to grow cane, and tree planting and maintenance costs. The former is estimated in terms of crop area lost and sugar cane gross margin ($/ha). Conditions of cane delivery to mills and price are laid down by the Central Sugar Cane Prices Board, based on the *Regulation of Sugar Cane Prices Act, 1962*. Quantity is measured as weight of cane delivered to mills; quality is expressed as percentage of pure sucrose (commercial cane sugar, CCS). The raw sugar price received by the miller is converted into a gross value for cane of varying qualities according to the cane-payment formula on the basis of recoverable sugar content. Once the CCS of a sample of cane and the 'coefficient of work' of a mill (indicating the quantity of raw sugar derived from the CCS used) are known, the raw sugar value can be expressed by the formula (Wegener *et al.*, 1990):

$$V = 0.009\,P\,(CCS - 4) + K$$

where V is price paid to the grower for cane ($/tonne),
P is the value of '94 net titre' sugar,
CCS - 4 is a factor originally designed to allocate proceeds 2/3 to growers and 1/3 to millers when the CCS was 12% and the mill coefficient of work was 90, and the amount of sugar the mill extracts is about 90%, and
K is a constant.

The final pool price of sugarcane in Queensland for the 1996 season was used, of $325/tonne (Rutledge, 1997). A long-term seasonal average CCS for North Queensland is approximately 12.5% and the value of K is set at 0.578. Therefore, the gross value (GV) of cane to the farmer is estimated as:

$$GV = \$0.009 \times 325 \times (12.5-4) + 0.578\ /\ \text{tonne} = \$25.44\ /\ \text{tonne}$$

The variable cost of growing and harvesting sugarcane is approximately $15 per tonne of cane, leaving a margin of $10.44/tonne. The approximate yield in the Scheu Creek catchment is 90 tonnes/ha, hence the gross margin is $940/ha. The opportunity cost of land under each option, estimated by multiplying the net area lost by this gross margin, is reported in Table 12.2.

Table 12.2 Estimated costs of the four riparian revegetation options along Scheu Creek

Cost item	Option			
	A	B	C	D
Land opportunity cost ($/yr)	6,298	11,092	9,118	18,800
Site preparation,tree planting ($)	171,520	302,080	248,320	512,000
Maintenance cost ($)	3,430	6,041	4,966	10,240

Bell (1996) reported estimated site preparation, tree planting and maintenance costs of $12,000/ha. According to NQJB (1997), costs of tree planting in the Johnstone Shire have been $25,600/ha; these costs appear to have been carefully estimated and are used in the present analysis. An allowance at 2% of these costs is made for maintenance costs in Years 1 and 2. Aggregate cost estimates are presented in Table 12.2.

12.5 BENEFITS OF RIPARIAN REVEGETATION IN SCHEU CREEK CATCHMENT

Various economic benefits have been identified for riparian revegetation:

Reduction in cost of rodenticides and weedicides. Two species of native rodent, the cane field rat (*Rattus sordidus*) and the climbing rat or grassland melomys (*Melomys burtoni*) cause damage to sugarcane in the Queensland wet tropics (Brodie, 1997a). The climbing rat is common in areas heavily vegetated with grasses and weeds, and especially near watercourses. Rats in these areas are a breeding nucleus for populations that colonise and damage crops, attacking plants both below and above ground (Agnew, 1997). Removing rat-harbouring areas reduces the potential for rodent activity (Brodie, 1997a). Traditional rat control methods involve baiting, combined with spraying weedicides on the perimeters of cane fields. Spraying has the side effect of increasing watercourse chemical residue levels. Canegrowers (1997) estimate a saving associated with tree planting along watercourses of $154/ha/year from reduction in use of chemicals and labour. It is assumed that the farmers spray a 2m wide strip to keep the creek open to water flow. According to GIS estimation, the total length of Scheu Creek is 13.3km, and the reduced area of spraying is 2.66ha. According to Brodie (1997b), cost savings in the four revegetation options would not differ. Table 12.4 presents total benefits due to reduction in cost of rodenticides and weedicides under Options A to D.

Reduction in rat damage cost. Rat damage directly reduces yield, and bitten cane is more prone to attack by bacteria, fungi, and insects including the sugarcane weevil. This can result in a 20% loss of CCS; a loss of 30% can result from a combination of lost tonnage and reduced CCS (Agnew, 1997). However, Clark (1997) has suggested that rat control will only provide a 5% increase in cane yield (a yield increase 4.5 t/ha) for Scheu Creek and this figure is adopted here. Quick growing trees can form a closed canopy within two years in the wet tropics, under which light reduction suppresses grass growth and reduces rat habitat. Rat numbers and rodent damage around Tully and Babinda in the North Queensland have fallen by 80% and 100%

respectively, where trees have been planted in riparian areas (Canegrowers, 1997). This report also claims that revegetation encourages owls into the area and individual families of owls can consume about 2,000 rats a year. A 10% decrease in loss of cane from rats has been claimed for North Queensland, from growing trees in riparian areas and encouraging owls by providing nesting boxes (ABC TV, 1996). This report further claimed that one owl could eat an average of eight rats per night. According to Canegrowers (1997), elimination of rat damage will increase cane harvest by 50% in the 20m of cane adjacent to the riparian tree buffer. Adopting this figure, the total area with reduced rat damage is 13.3km x 20m x 2 sides, or 53ha. Assuming a 5% reduction in damage (i.e. yield increase of 4.5 t/ha) and a gross return from cane net of harvesting costs of $10.44/ha, the benefit from reduction in rat damage is estimated as 53ha x 4.5t/ha x $10.44/ha or $2,490, as indicated in Table 12.4.

Reduction in flood damage cost on-farm. Replacement of Para Grass with trees can be expected to increase streamflow and reduce crop loss due to flood damage. The total area of the Scheu Creek catchment is approximately 651ha (Allan, 1998), of which an estimated 586ha is under cane. Following Canegrowers (1997), it is assumed that due to the reduction in flood damage, there will be a 1% increase in overall cane yield once every 10 years over this area. The increase is taken to be in year 5, 15 and 25. Assuming a yield of 90t/ha, the benefits in the whole catchment due to flood reduction are 90t/ha x 1% x 586ha x $19/tonne or about $10,000, as indicated in Table 12.4. It is assumed that these amounts will not vary between revegetation options.

Reduction in off-site damage costs. Transport and deposition of silt from the Scheu Creek catchment in periods of intense rain potentially causes water pollution, road damage and flooding of cane in the lower catchment. Day (1997) has provided subjective estimates of erosion related damage to the roads in the Scheu Creek catchment once every five years as $10,000. Prove (1997) made subjective estimates of the reduction in road maintenance cost due to riparian revegetation options A to D of 10%, 15%, 13% and 15% respectively. To estimate the reduction in road maintenance costs, Day's (1997) estimates of road damage costs and Prove's (1997) estimates of reduction in these costs due to riparian revegetation options are adopted. The estimated benefits under options A to D on a once-in-a-five-years basis are $1,000, $1,150, $1,300 and $1,150 respectively (Table 12.4). No evidence could be found of recent dredging activities in the Johnstone River, therefore no cost is attributed to dredging.

CO_2 sequestration benefits due to tree planting and growing. Trees planted along Scheu Creek would sequester CO_2 and make some contribution to reduction of greenhouse gases. Derivation of carbon sequestration benefits, based on approaches of van Kooten *et al.* (1995) and Thompson *et al.* (1997), and estimates by Eono (1997) for native tree species in North Queensland, is reported elsewhere (Qureshi, 1999); estimates are summarised in Table 12.3.

Table 12.3 Estimated MAI due to four riparian revegetation options

Option	A	B	C	D
Area planted (ha)	6.7	11.8	9.7	20
Average annual timber production (m^3)	67	118	97	200
CO_2 sequestration benefits ($/year)	335	590	485	1,000

Reduction in downstream flooding. A reduction in severity of downstream flooding can be expected, but this would probably be relatively small because Scheu Creek is small relative to the overall catchment and the reduction in runoff rate from the riparian buffer option relative to current vegetation may not be great. No allowance has been made of this benefit.

Impact on streambank stability and water quality. Riparian revegetation options can be expected to improve streambank stability and water quality and improve the riverine environment. The estimation of these benefits is difficult and no attempt is made to quantify them in dollar terms. The cost of streambank slumping could be measured in terms of the earthworks costs of repairs. The water of the Scheu Creek is not used for human consumption, therefore no reduction in cost of water treatment is involved, and no benefits are gained due to revegetation in this regard.

Reduced impact on the Great Barrier Reef. In periods of intense rainfall, sediments and nitrates originating on farms are transported by the Johnstone River and discharged into the ocean. Should these cause damage to the Great Barrier Reef (GBR), reef health, fishing and tourism could be adversely affected. Moss *et al.* (1992) estimated that on average 15 M tonnes of sediment, 77,000 tonnes of nitrogen and 11,000 tonnes of phosphorus are discharged annually from Queensland's coastal catchments, and the bulk of these are discharged to the GBR lagoon. There has been concern that runoff from agricultural land in coastal North Queensland and transported sediment causes or contributes to outbreaks of algae and loss of coral (Bell and Elmetri, 1995). However, whether sediment deposition and associated transport of agrochemicals from cane lands is causing significant damage to the Great Barrier Reef is still conjectural, and no benefits have been attributed to riparian revegetation in this context.

Healthier instream ecosystem. Riparian trees and shrubs could help maintain good habitat for aquatic animals including insects, and increase numbers of fish that feed on them in the creek. Estimation of these benefits in dollar terms is difficult because of the complex ecological systems and no values have been placed on them in this study.

Table 12.4 Estimated benefits of riparian revegetation

Benefit item	Option			
	A	B	C	D
Reduction in cost of rodenticides and weedicides ($/year)	40	409	409	409
Reduction in rat damage cost ($/year)	2,49	2,490	2,490	2,490
Reduction in flood damage costs ($/once-in-10 years)	10,00	10,000	10,000	10,000
Reduction in off-site damage costs ($/once-in-five years)	1,00	1,150	1,300	1,150
CO_2 sequestration benefit ($/year)	335	590	485	1,000

Potential returns from timber extraction. If high-value cabinet timber species are planted in riparian areas, there is the potential for future selective logging. Intensive timber harvesting would defeat the purpose of developing natural habitat, and would not be beneficial for the riverine environment. As well, logging activity along watercourses is restricted under the *Water Resource Act 1993*. An argument in favour of tree harvesting is that if trees are left to die, they may fall over and disturb a large soil mass around the roots. Also, if some of the original species become dominant, there may be a need to alter the species balance. Some very high value timber species grow well along watercourses. Due to the high uncertainty about harvesting, no allowance has been made for timber revenue in the economic analysis.

12.6 DISCOUNTED CASH FLOW ANALYSIS REVEGETATION POLICY OPTIONS

Microsoft Excel software is used to determine net present values for the four revegetation options. Costs and benefits under each of the categories have been derived for each year. The annual net cash flow is obtained as

$$NCF_t = -LOC_t - TPC_t + SRWPC_t + RRDC_t + RFDC_t + ROSDC_t + CSB_t$$

where NCF = net cash flow
LOC = land opportunity cost
TPC = tree planting and maintenance costs
SRWPC = savings in rodenticide and weedicide prevention costs
RRDC = reduction in rat damage costs
RFDC = reduction in flood damage costs
ROSDC = reduction in off-site damage costs
CSB = CO_2 sequestration benefits
t = time period

The costs and benefits derived above are entered in a spreadsheet (Table 12.5) to estimate present values of revegetation options. The analysis is carried out from the society's point of view. The estimated social NPVs of the four riparian revegetation options (A to D) are summarised in Table 12.8. None of the four options has a positive NPV, and the highest net cost is obtained for Option D (i.e. $718,000), and the lowest for Option A ($205,000). It is to be noted that some benefits of riparian revegetation have not been quantified here, e.g. impact on water quality, bank stability, impact on the Great Barrier Reef. The estimated NPVs indicate how high these other values would need to be to justify the revegetation options on overall economic grounds.

Because of the high degree of uncertainty about the costs and benefits of the riparian revegetation options, it is useful to perform sensitivity analysis with variations to a number of parameters. This provides information about the effects of errors in parameter estimates on social NPV. Sensitivity analysis has been performed with respect to three major parameters, discount rate, tree planting costs and reduction in rat damage costs (Tables 12.6 and 12.7). Table 12.6 indicates that changing the discount rate has little effect on NPV. While there are considerable up-front costs, foregone crop revenue is spread throughout the planning horizon. A change in discount rate does not change the ranking of the revegetation options. While NPVs change with variations in other parameter levels, the ranking of four revegetation policy options does not change (Table 12.7).

Costs and benefits may also be examined from a private landholder's point of view. Here, changes in tax payments are taken into account, while offsite benefits (e.g. reduction in road damage costs, carbon sequestration benefits) are excluded. Costs of tree planting and maintenance are taken as fully deductible from farm income for tax purposes, while additional income would be taxed at the same marginal rate. The estimated private and social NPVs of the four riparian revegetation options (A to D) are summarised in Table 12.8. Again, none of the estimates are positive, though absolute (private) values are small compared to the social NPVs.

Table 12.5 Social costs and benefits of riparian revegetation, Option A

Year	Opportunity cost of land ($)	Tree planting costs ($)	Reduction in rodenticide costs ($)	Reduction in rat damage costs ($)	Reduction in flood damage costs ($)	Reduction in offsite damage costs ($)	CO_2 seqn. benefits ($)	Net benefits ($)
0	6,298	171,520						-177,818
1–2	6,298	3,430						-9,728
3–4	6,298							-6,298
5	6,298		409	2,490		1,000	335	-2,064
6–9	6,298		409	2,490			335	-3,064
10	6,298		409	2,490	10,000	1,000	335	7,936
11–14	6,298		409	2,490			335	-3,064
15	6,298		409	2,490		1,000	335	-2,064
16–19	6,298		409	2,490			335	-3,064
20	6,298		409	2,490	10,000	1,000	335	7,936
21–24	6,298		409	2,490			335	-3,064
25	6,298		409	2,490		1,000	335	-2,064
26–29	6,298		409	2,490			335	-3,064
30	6,298		409	2,490	10,000	1,000	335	7,936
NPV (@ 7% discount rate) =								-221,856

12.7 IMPLEMENTATION OF REVEGETATION OPTIONS

To implement revegetation options, resource managers need to take account of the social and private costs and benefits[1] of alternatives, attitudes and preferences of the various stakeholder groups, and relevant legislation and regulations. It may well be that the benefits of riparian revegetation exceed the costs, when benefit classes not included in the above analysis are taken into consideration. However, it is apparent that the costs will weigh most heavily on farmers, while benefits will be enjoyed by a wider constituency of stakeholders. The question thus arises as to whether the landholders should be expected to meet the revegetation costs since the land should not have been cleared in the first place (polluter pays principle) or the wider

[1] The multicriteria analysis conducted in this study allows other benefits for which monetary estimates cannot be made with acceptable confidence to be factored into the overall evaluation in quantitative terms.

community should compensate the farmers for revegetation costs (beneficiary compensates principle). Much of the riparian clearing would have taken place several generations ago. Some of the farmers would be of pioneering families while others would have purchased developed land.

Table 12.6 Effects of altering discount rates on NPV ($1,000)

Revegetation option	Discount rate (net present value) ($1,000)		
	4%	7%	10%
A	-231	-222	-215
B	-447	-417	-398
C	-360	-337	-323
D	-803	-736	-694

Table 12.7 Effect of a change in tree planting and rat damage costs on NPV ($1,000)

Option	Base	Tree planting cost		Reduction in rat damage costs	
	NPV	30% increase	30% decrease	30% increase	30% decrease
A	-222	-275	-169	-215	-229
B	-417	-511	-260	-411	-424
C	-337	-414	-233	-331	-344
D	-736	-757	-715	-729	-743

Table 12.8 Net present values of revegetation policy options

Option	A	B	C	D
NPV ($1,000), social	-222	-399	-319	-718
NPV ($1,000), private landholder	-160	-295	-240	-515

While there are various Queensland Acts to protect catchment resources, none provide the power for resource managers to force landholders to revegetate riparian areas. For equity and practical reasons, it would appear desirable to provide taxpayer support to promote tree planting along watercourses, by measures such as extension and moral persuasion, technical advice, provision of tree seedlings, reimbursement of part of the expenditure, and encouragement of tree planting initiatives by Landcare or other community groups. Legislative force or local government bylaws would seem a last resort if these resources were unsuccessful.

12.8 SUMMARY

Discounted cash flow analysis of riparian revegetation policy options for Scheu Creek catchment allows most costs and some of the benefits to be quantified in present day dollar terms. Naturally, the wider the riparian buffer, the greater the net cost. Even when the uncertain parameters are varied over wide ranges, NPVs remain negative for the cost and benefit categories included in the analysis. Inclusion of additional benefits may well justify riparian revegetation, in which case attention needs to be focused on cost-effective implementation methods. Further benefits in terms of revegetation on land and watercourse stability, impact on groundwater and surface water quality and land and stream habitat, and impact on the Great Barrier Reef can be taken into account in a wide-ranging multicriteria analysis.

REFERENCES

ABC TV (1996), Landline TV Programme, Australian Broadcasting Corporation.

Agnew, J.R. (1997), 'Australian Sugar Cane Pests', Tully: Bureau of Sugar Experiment Stations.

Allan, B. (1998), Personal Communications, Technical Officer (Drafting), South Johnstone: Department of Natural Resources.

Bell, P.R.F. and Elmetri, I. (1995), 'Ecological Indicators of Large-Scale Eutrophication in the Great Barrier Reef Lagoon', *Ambio*, **24**(4), pp. 208–215.

Bell, R. (1996), 'Johnstone River Catchment Revegetation Strategy', Innisfail: Johnstone River Catchment Management Association Inc., Ch. 6, pp. 41–43.

Brodie, A. (1997a), 'Trees for Rat Control', in proceedings of forum on *Management of Riparian Lands in the Wet Tropics Region: Integrating Policy and Research with Stakeholder Needs*, K. Johnson (ed.), Innisfail: Johnstone River Catchment Management Association Inc..

Brodie, A. (1997b), Personal Communication, Tully: Bureau of Sugar Experiment Stations.

Bunn, S. E. and Davies, P.E. (1996), 'The Influence of Riparian Vegetation on Stream Ecosystem Health in the Wet Tropics', in *Integrating Policy and Research with Stakeholder Needs*, Innisfail Johnstone River Catchment Management, pp. 38–42.

Canegrowers (1997), *Riparian Management: Is there a Rat in Your Hip Pocket*, Brisbane: Canegrowers Association.

Clark, N. (1997), Personal Communication, Innisfail: South Johnstone Cane Pest and Productivity Board.

Day, S. (1997), Personal Communication, Engineer, Cairns: Main Roads Department.

Eono, J-C. (1997), 'Carbon Sequestration in the Community Rainforest Reforestation Programme (CRRP)', in *Agricultural and Resources Economics Society Conference*, Gold Coast.

Goosam, S. and Tucker, N.I.J. (1995), *Repairing the Rainforest: Theory and*

Practice of Rainforest Re-Establishement in North Queensland's Wet Tropics, Cairns: Cassowary Publications.

Herron, N. (1996), 'A Review of Water Flow Pathways Through the Riparian Zone', in I. Rutherfurd and M. Walker (eds), *Proceedings of First National Conference on Stream Management in Australia*, Canberra: Cooperative Research Centre for Catchment Hydrology, pp. 197–202.

Moss, A.J., Rayment, G.E., Reilly, N. and Best, E.K. (1992), '*A Preliminary Assessment of Sediment and Nutrient Exports from Queensland Coastal Catchments*', Report to the Queensland Department of Environment and Heritage and Department of Primary Industries, Brisbane.

NQJB (North Queensland Afforestation Programme Joint Board) (1997), *Wet Tropics Tree Planting Scheme Operations Report 1/7/96 – 30/6/97*, Internal Unpublished Report of North Queensland Joint Board, Cairns.

Prove, B. (1997), Personal Communication, South Johnstone: Centre for Wet Tropics Agriculture, Department of Natural Resources.

Qureshi, M.E. (1999), Development and Implementation of a Decision Support Process for Sustainable Catchment Management, PhD thesis (unpublished), The University of Queensland, Brisbane.

Raine, A.W. and Gardiner, J.N. (1995), *RIVERCARE: Guidelines for Ecologically Sustainable Management of Rivers and Riparian Vegetation*, Occasional Paper No. 03/95, Canberra: Land and Water Resources Research and Development Corporation.

Russell, I.W., Izac, A-M.N. and Cramb, R.A. (1990), *Towards an Evaluation of the Off-site Costs of Soil Erosion in Queensland*, Agricultural Economics Discussion Paper Series, 2/90, Brisbane: Department of Agriculture, The University Queensland, pp. 1–91.

Rutledge, D. (1997), 'Sugar Cops it Sweet Amid Asian Currency Crisis: The Asian Financial Crisis is Casting Pall Over an Otherwise Buoyant Outlook for the Sugar Industry', reported by Rhyll Cronin, Brisbane: *Courier Mail*, Wednesday 29 December, 1997, pp. 13-14.

Thompson, W.A., van Kooten, G.C. and Vertinsky, I. (1997), Assessing Timber and Non-Timber Values in Forestry Using a General Equilibrium Framework, in R.A. Sedjo, R.N. Sampson and J. Wisniewski (eds), *Economics of Carbon Sequestration in Forestry*, London: Lewis Publishers, pp. S351–S364.

van Kooten, G.C., Binkley, C.S. and Delcourt, G. (1995), 'Effects of Carbon Taxes and Subsidies on Optimal Forest Rotation Age and Supply of Carbon Services', *American Journal of Agricultural Economics*, **77**(5), pp. 365–374.

Wegener, *et al.* (1990), 'Agro-Industrial Crops', in R.L. Reid (ed.), *The Manual of Australian Agriculture*, Sydney: Butterworths, Ch. 6, p. 121.

13. The Inclusion of Non-Wood Benefits in Reporting Systems of Forest Enterprises

K.F. Herbohn, S.R. Harrison and J.L. Herbohn

Traditional accounting procedures for forestry enterprises fail to measure non-wood benefits. A more comprehensive accounting system would better discharge the reporting responsibilities of these agencies. Developing credible measures of non-wood benefits is a far from simple task. However, progress is being made in this endeavour, as illustrated by an example for farm forestry in tropical North Queensland, Australia.

13.1 INTRODUCTION

Participation in forestry by individuals and organisations is not motivated solely by the generation of commercial returns from the sale of harvested timber (i.e. wood benefits). For example, a survey of individual landholders' motivations for participation in the Queensland Joint Venture Scheme by Harrison *et al.* (1999) revealed both economic and environmental aims. The provision of technical expertise in growing trees, the opportunity to generate financial returns and the input of capital by the Department of Primary Industries (DPI) were found to be of importance to landholders, but so too are non-wood benefits such as landscape amenity, conservation benefits, and land rehabilitation. Organizations such as the DPI also acknowledge the multidimensional nature of an investment of time and resources in forestry. 'It is now generally recognised that Australia's forests are a valuable resource; ecologically, socially, aesthetically and economically. However, the appreciation of these difference roles has meant that forests are now managed for a number of different objectives . . .' (QDPI, 2000).

Accounting systems are designed for communication of relevant information about an organization's performance and financial position to interested individual stakeholders and stakeholder organizations. Yet, traditional accounting systems are ill equipped to identify, measure and communicate information on the non-wood benefits accruing to a reporting entity during the period. Non-wood benefits are excluded from traditional accounting reports because these reports are prepared from only a partial information set, which does not include any unpriced outcomes and events. Much accounting wisdom is drawn from neoclassical economics and so any event that is not supported by a market transaction and therefore a market price, is excluded from accounting systems. Consequently, while stakeholders may support an investment in forestry because it has non-wood benefits such as land protection and the provision of wildlife habitats, accounting reports will only inform them about the wood benefits resulting from the harvest and sale of timber. This chapter provides an overview of the current state of reporting non-wood benefits amongst Australian forestry entities, and suggests an approach to incorporating non-wood benefits into traditional accounting reports that has been experimented with in Australia and Italy.

13.2 REPORTING NON-WOOD BENEFITS WITHIN AN AUSTRALIAN ACCOUNTING FRAMEWORK

Over the last decade, the accounting sector within Australia has undergone significant change. This has translated into changes for reporting entities generally and within specific sectors such as superannuation, insurance and forestry. The main changes for forestry agencies that prepare accounting reports are a broadening of the reporting net, and a new accounting regulation on self-generating and regenerating assets. These developments are briefly reviewed to provide a background on attempts to incorporate non-wood benefits into accounting frameworks.

The Current Forestry Accounting Framework in Australia

Generally, accounting regulators from the Australian Accounting Standards Board (AASB) and the Public Sector Accounting Standards Board (PSASB) require an entity to prepare general purpose accounting reports, if users exist who are dependent on the reports to make resource allocation decisions. These reports usually comprise a balance sheet (statement of financial position), profit and loss statement (statement of financial performance) and a cash flows statement. User groups may include employees, shareholders, consumer groups, lobby groups (e.g. environmental groups), lenders and organisations with an oversight function. Many organisations such as exempt proprietary companies, trusts, sole traders and partnerships have been deemed not to have such dependent user groups, and so have not been required to prepare accounting reports.

Recently, Australian accounting regulators have broadened the general reporting requirement beyond the existence of dependent users (Herbohn and Herbohn, 1998). Corporations Law has been amended to include a size test for reporting based on gross revenue, gross assets and the number of employees, which means that companies above a certain size fall within the reporting net. The implication for forestry entities is that many mid-sized privately-held forestry companies may now have to prepare general purpose accounting reports. As part of recent moves to increase the accountability and efficiency of the public sector, the reporting net has also been extended to include public sector organizations. This has meant that public sector organizations such as State Forest Services must now prepare accounting reports using the same accounting principles as large commercial private sector organizations.

Another change for forestry organizations is the introduction of an applicable accounting standard – Accounting Standard AASB 1037 *Self-generating and regenerating assets* – in August 1998, with an operational date of June 2000. AASB 1037 deals with forestry, livestock and other self-generating asset industries (e.g. consumable crops, orchards, vineyards, aquaculture). However, only recommendations relating to forestry assets are considered here. Not surprisingly, AASB 1037 deals only with how to report on the wood benefits from an investment of time and resources into forestry. For example, the standard stipulates the ways in which forest assets are to be valued (i.e. at net market value); forest assets are to be reported in the balance sheet (i.e. separately as a sub-category of non-current assets); and changes in the value of forest assets prior to harvest, due to changes in timber volume, timber quality and timber prices, are to be recognised (i.e. as revenue or expense in the profit and loss statement).

Regulatory Framework for Reporting of Non-Wood Benefits in Australia

When developing any accounting policy, the AASB and PSASB try to ensure a transparent process with the opportunity for debate amongst interested constituents. As part of this process, a discussion paper is commissioned to provide a broad discussion of the issues and potentially appropriate accounting techniques – in this case Discussion Paper No. 23 (DP 23) *Accounting for self-generating and regenerating assets* which was released in May 1995. The next step is the release of a draft policy or exposure draft that supersedes any discussion paper. In the case of forestry, the document was Exposure Draft No. 83 (ED 83) *Self-generating and regenerating assets* released in August 1997.

During the evolution of AASB 1037, there was an attempt in DP 23 to require some form of disclosure on the non-wood benefits of forestry activities. It was noted that financial measurements of some natural resources, including forests, may not be sufficiently reliable or relevant because of measurement difficulties, and so non-financial disclosures may be more appropriate (e.g. descriptions of qualities of assets, insurance cover for losses, geographic segment information). The accounting profession was also encouraged to be more active in promoting accountability relating to the environment, because of the emphasis placed on economically sustainable development of forests by governments and society in general. However, any reference to these disclosures was deleted from ED 83 and then AASB 1037 with no justification. There is currently no requirement in Australia to report on the non-wood benefits resulting from forestry operations.

Voluntary Reporting of Non-Wood Benefits in Australia

Since at least the 1970s, accountants have acknowledged the constraints of a traditional reporting system and have grappled with the possibility of 'attempting to incorporate the social and environmental dimensions of organizational life into the accounts and accountings that are produced by economic entities (variously) as performance reports, stewardship documents, decision support information and so on ' (Gray, 1994, pp. 5–6). These efforts to extend the scope of traditional accounting have not had regulatory backing. Instead, researchers and organizations in America, Australasia and Europe have undertaken experimentation with traditional reporting on an ad hoc, voluntary basis, and in some cases the reporting has been an advertising, image enhancing exercise devoid of any information content (Harte and Owen, 1992; Gray *et al.*, 1995).

The motivations of entities undertaking environmental and social reporting experimentation have been mixed. Some entities have undertaken this experimentation with altruistic aims concerned with discharging full accountability to their stakeholders in society (Freeman, 1984; Roberts, 1991). In contrast, the motivations of other organizations have been self-serving and there has been debate on whether organizations can accept any social and environmental responsibilities because of commercial pressures to maximise financial returns to owners (Friedman, 1962). Instead, these organizations have been concerned with the positive impacts of social and environmental disclosures on share prices, and the political benefits of changed perceptions of government and employees (Ullman, 1985; Gray *et al.*, 1993).

The trends observed in surveys of environmental and social accounting experiments – ad hoc experimentation, and mixed motivations – are also evident in attempts within the Australian forestry sector to report on non-wood benefits. A survey by Herbohn *et al.* (1998) of environmental disclosure practices of public and private sector forestry organizations documented three main trends:

- First, environmental reporting by the public sector entities surveyed is more comprehensive than that of the private sector entities. Most of the public sector forestry entities incorporate environmental information into their annual reports. This information includes a mission statement and list of goals, descriptions of environmental activities, physical data on forests (e.g. number of species planted), and performance indicators showing progress toward stated goals (e.g. forest user survey results).
- Second, there is a range of disclosure practices adopted by private sector entities from no environmental disclosures to minimal disclosures of amounts spent on environmental activities, qualitative descriptions of environmental activities, and descriptions of environmental objectives.
- Third, while half of the private sector entities surveyed expanded their environmental disclosures (e.g. creation of a separate environmental section within the accounting report), the other half down-scaled such disclosures (e.g. discontinuance of a separate environmental section with the accounting report).

Clearly, there is scope for experimentation with environmental and social reporting in the Australian forestry sector to allow communication of information on non-wood benefits to interested stakeholders.

13.3 AN EXPERIMENTAL APPROACH TO REPORTING NON-WOOD BENEFITS BY FORESTRY ENTITIES

Accountants have experimented with preparing financial environmental accounts for at least three decades (Abt and Associates, 1972; Bebbington and Gray, 1997). These accounts attempt to place a financial value on externalities that is incorporated with other financial information into accounting reports. The underlying rationale is that externalities such as non-wood benefits will feature more prominently in decision-makers' priorities because they will be reflected in the 'bottom line' in the profit and loss statement, and so cannot be ignored. This approach attempts to overcome the situation whereby 'a country could exhaust its mineral resources, cut down its forests, erode its soil, pollute its aquifers, and hunt its wildlife to extinction, but measured income would not be affected as those assets disappeared' (Repetto *et al.*, 1989, p. 2).

One impediment to progress in development of financial environmental accounts has been the traditional argument that it is outside the scope of orthodox accounting. It is argued that accountants do not possess the requisite ecological and scientific knowledge, nor the measurement methodologies, to identify unpriced environmental benefits and costs, disclose them within accounting reports, and defend these estimates under legal challenge. There have even been concerns that the techniques of assigning a value to externalities such as benefits from the aesthetic pleasure provided by an architecturally designed office block, or to costs such as air pollution, are not fully developed (Gray *et al.*, 1987). The validity of these arguments has been called into question by developments in the discipline of environmental and natural resource economics.

Environmental economists have developed techniques to measure unpriced benefits from public goods such as water, wildlife, air quality and aesthetics. The conceptual framework usually adopted is that of total economic value (TEV), comprising use value and non-use (option and existence) value (Pearce and Moran, 1994). A good deal of research is being devoted to estimating TEV and its individual components. The pace of development has been such that Deacon *et al.* (1998, p. 388) have observed 'it is doubtful that anyone could have anticipated the explosion of research on non-market valuation that has taken place over the last 25 years'. Methods investigated to date include the stated preference techniques of contingent valuation, contingent rating, contingent ranking and choice modelling (e.g. Sterner and van den Bergh, 1998; Hanley *et al.*, 1998), and surrogate market techniques such as the travel cost and hedonic price methods. As well, there is considerable interest in the expedient of benefits transfers which involve transferring existing estimates of unpriced impacts to a new study (Boyle and Bergstrom, as reported in Kirchloff *et al.*, 1997). Space precludes further discussion, but a contemporary survey of progress and challenges in non-market valuation is provided by Harrison (1999).

Several accountants (e.g. Milne, 1991, 1996; Palmer, 1998) and organizations (e.g. Earth Sanctuaries, Dow Chemical, Ontario Hydro) have begun to explore how economic valuation techniques might be incorporated into an accounting framework. Generally, the discussion and experimentation has centred on two possible applications of the measurement techniques from natural resource and environmental economics. First, estimates of unpriced environmental and social impacts could be incorporated into internal costing frameworks (e.g. Dow Chemical, Ontario Hydro), so that internal decision-makers have a more complete information set with which to evaluate investment projects, set objectives, evaluate organizational performance and develop pricing policies (Milne, 1991, 1996). Second, estimates of unpriced environmental and social benefits and costs could be incorporated into external accounting reports that are used to communicate with an entity's stakeholder groups (Palmer, 1998). Stakeholder groups would have a more complete information set that would allow more informed decisions on the organization's overall financial, environmental and social performance.

While experimentation with internal costing using measurement techniques from environmental economics is a fruitful area, the focus of the chapter is on external reports used to communicate organizational performance to stakeholders. To date, there have been at least two reporting experiments involving forestry organizations that have been aimed at using environmental economics valuation techniques to communicate information on non-wood benefits to external stakeholders. These experiments are briefly outlined below.

13.4 REPORTING NON-WOOD BENEFITS OF COMMUNITY FORESTRY IN NORTH QUEENSLAND

An extended reporting framework has been proposed for the Community Rainforest Reforestation Program (CRRP) operating in North Queensland, Australia. The CRRP was developed in response to World Heritage listing of previously logged forests and was managed by a committee with representation from each tier of government. The programme had four objectives, stated to be of equal importance, that are listed below (CRRP Management Committee, 1996): develop a private plantation base for a sustainable timber industry; address the problem of land degradation following extensive inappropriate clearing; provide improved stream quality by establishing vegetative buffers along rivers and streams; and train a workforce to support the long-term practice of rainforest plantation establishment. Consequently, the CRRP has many different stakeholders including landholders, local government, environmental groups, and funding bodies that are interested in information on wood *and* non-wood benefits.

A reporting system was proposed to assist in the communication of wood and non-wood benefits to stakeholders in the CRRP, a detailed discussion of which is provided in Herbohn *et al.* (2000). This reporting framework is based on the original functions of the CRRP, which have been modified over time. An inventory of environmental impacts of the CRRP was developed, resource or environmental changes were matched to appropriate valuation techniques, and this information was incorporated into an environmental disclosure format drawn from Dierkes and Preston (1977) and Gray *et al.* (1993). An extract from the proposed reporting framework is presented in Table 13.1. On the basis of this information, the summary evaluation is drawn that

Table13.1 Extract from proposed special purpose environmental report for the CRRP for Water quality (CRRP objective 3)

Environmental impact dimension	CRRP data		Comparison norms		
	Program /Policy	Experience	Industry	Commun-ity	Other
1. Protection of water quality	X	X Scientific measurement of agricultural runoff concentrations in water downstream from CRRP sponsored plantations.	X	X	Results of scientific research into numbers (and changes therein) of animals and plants supported by waterways in regions with CRRP sponsored tree plantations. Levels of agricultural runoff identified as a problem by Australian scientists.
2. Reduced sedimentation of water courses	X	Scientific measurement of particle concentration in a water sample downstream from CRRP plantings.	X	X	Generally acceptable levels among Australian scientists of particle concentration within water. Average number of times per annum (and changes therein since establishment of the CRRP) that local port authorities dredge harbours in CRRP plantation regions.
3. Protection against stream bank erosion	Sensitive logging guidelines	X	X	X	X
4. Increased opportunities for recreation (e.g. fishing) on waterways	X	Environmental economic valuation of increased opportunities for recreation on waterways ($)	X	X	X
5. Improved aesthetics of waterways	X	Environmental economic valuation of improved aesthetics ($)	X	X	X

Note: X indicates where disclosures will be made when information becomes available. For example, the industry and community columns will outline accepted industry or community standards for environmental performance.

Water quality appears to have improved in regions with CRRP sponsored plantations, with supporting evidence from scientists and local port authorities. Also, improvements in water quality appear to have had a positive impact on the value of waterway aesthetics and perceived recreation opportunities along waterways.

The information is not disclosed in traditional accounting reports (e.g. balance sheet, profit and loss statement). Instead, a special purpose environmental report is recommended that is divided into five sections based on the four objectives of the CRRP, and the perceptions of stakeholders identified in surveys (e.g. Creighton and Sexton, 1996; Harrison *et al.*, 1998) that are summarised by a fifth objective labelled 'other'. Each section contains environmental information such as financial estimates of unpriced non-wood benefits (e.g. the value of improved recreation opportunities along waterways), quantitative data (e.g. number of species planted), and qualitative information (e.g. descriptions of improved farm aesthetics). Comparison norms from industry, the community and other (e.g. State Government guidelines) are also included to assist statement users to assess the CRRP's overall performance.

Traditional accounting reports are usually subject to review by an independent auditor who verifies that an organisation's report is in accordance with Corporations Law, which includes providing a true and fair view of the entity's financial position and performance, and complying with Accounting Standards. Due to the cost involved in an external audit, it was recommended that the CRRP's special purpose environmental report be reviewed by a well-respected critical journalist or a member of the forestry sector, and that the subsequent critique be published with the report.

13.5 REPORTING NON-WOOD BENEFITS OF A PUBLIC FOREST IN NORTHERN ITALY

Merlo and Jobstl (1999) have proposed a step-wise approach to improve forestry accounting by including unpriced environmental values. Their system has been tested on the Cansiglio multi-purpose public forest located in the Veneto region in north-eastern Italy. The underlying motivation for Merlo's and Jobstl's (1999) experiment was that even though there is a clear demand for information on non-market values relating to issues such as landscape quality and free recreation, traditional forestry accounting reporting in Europe does not incorporate such information. They argue that forestry operations are in a unique situation because much of the impetus for environmental accounting has stemmed from a desire to make organizations that are depleting and degrading natural resources accountable for the negative effects of their activities. Yet, forestry operations have the potential to address problems of timber depletion, increase biodiversity and improve landscape quality, and so in many cases may result in net environmental benefits to society. Their aim is to reflect the value of non-wood benefits from forestry in financial reports.

Merlo and Jobstl (1999) recommend the internalisation of non-wood benefits in a step-wise process, with each of the four steps resulting in more comprehensive environmental reporting. They view the process as moving from a 'light green' environmental reporting option (step one) through to a 'dark green' environmental reporting position (step four).

Step 1. Modification of the balance sheet to report assets, liabilities and capital, with all valuations at market value. The profit and loss was also modified where possible to distinguish between revenues and expenses from traditional activities (i.e. timber production), and the new production of environmental products and services (e.g. mushroom picking, guided walking tours). Table 13.2 contains an overview of the environmental balance sheet format adopted for the Cansiglio forest and Table 13.3 contains on overview of the profit and loss statement prepared for the Cansiglio forest.

Step 2. Identification and reclassification of expenditure undertaken to conserve or improve the environment to allow separate disclosure in the profit and loss statement. (See table 13.3 section E and F of the profit and loss statement for this separate disclosure.)

Step 3. Identification of hidden environmental assets, and liabilities. For example, it was identified that the Cansiglio forest gave rise to a non-marketable increase in growing tree stock. This increase was valued and recorded as an increase in assets and an increase in profit. There was also an element of risk relating to the possibility of disease affecting artificial stands due to the lack of biodiversity and natural protection was also identified. It was recognised as a depreciation charge against profit, and disclosed as a liability in the balance sheet.

Step 4. Reporting unpriced values of non-timber benefits through preparation of addenda for the balance sheet and profit and loss statement. The addendum to the balance sheet lists physical quantifies of resources (i.e. factors employed to produce public services) and liabilities (public goods borrowed from society). The addendum to the profit and loss statement reports profit adjusted for monetary estimates of unpriced benefits and off-site flows of services, e.g. the value of mushroom picking in the forest without permits was valued using the travel cost and contingent value methods (Table 13.3).

Table 13.2 Economic environmental balance sheet of Cansiglio Forest 1992–1994, from by Merlo and Jobstl, 1999 (average values, 1,000 ECU)

Assets		Liabilities	
Net assets	10,236.5	Creditors	17.5
Non-marketable fixed assets	35.0	Provision for pensions	105.0
(growing stock increase)		Capital and reserves	10,110.5
Raw materials, finished goods	40.0	of which	
Debtors	6.0	Net profit/loss for year	
Cash	5.5	adjusted for hidden	
		environmental values	0.5
Total net fixed assets	10,323.0	Total liabilities	10,323.0
ADDENDA			
Social Assets		Social Liabilities	
Protection forest ha 160	.....	Environmental debts	
(protection)		toward society	
Recreation forest ha 800	.....	and recreation forests,	
Wildlife (no. of species,	.....	wildlife, game, etc)	.....
quantity)		Net profit, loss for the	
Game (no. of head)	.....	year adjusted for non-	
		market environmental	.....

Merlo and Jobstl (1999) do not attempt full integration of environmental and financial information. Instead, they prepare what they term addenda to the balance sheet and the profit and loss statement for the following three reasons. First, integration of environmental and financial information is a radical option. Second, it would cause problems of valuation, integration and interpretation of information on the part of management, the authorities and the general public. Third, the use of addenda is more practical, easier to implement and assures the continuity of existing information systems. Details of the fourth step are provided below.

13.6 FUTURE CONSIDERATIONS IN REPORTING NON-WOOD BENEFITS

Environmental reporting experiments – such as that with the CRRP in Australia and the Cansiglio forest in Italy – are significant because they are a response to increased environmental awareness amongst stakeholders in society. This experimentation is aimed at improving communication with stakeholders by reporting on *all* aspects of a forestry operation to enable more informed decisions to be taken. For example, a reforestation scheme may be able to mount a stronger case to its funding bodies for continued resources if positive non-wood benefits can be quantified and included in reports.

Table 13.3 Profit and loss account of Cansiglio Forest 1992–1994, prepared by Merlo and Jobstl (1999) (average values in 1,000 ECU)

Profit and loss account of Consiglio Forest		
A) Revenues from ordinary activities		347.0
B) Expenditures for ordinary activities		351.0
RESULT FROM ORDINARY ACTIVITY (A-B)		-4.0
C) Revenues from recreation and environment		67.5
D) Expenditures for recreation and environment		721.5
RESULT FROM RECR. AND ENVIRONMENT (C-D)		-654.0
E) Financial expenditures and revenues		-4.5
F) Increase/decrease in market value of fixed assets		100.0
PROFIT/LOSS PRE-TAXES/COMPENSATIONS (A-B+C-D+E+F)		-563.0
Compensations (grants, subsidiaries and incentives)		533.0
G) NET PROFIT/LOSS FOR THE YEAR		-30.0
H) Risk depreciation of artificial stands [1]		-4.5
I) Non-marketable growing stock increase [2]		35.0
J) NET PROFIT/LOSS FOR THE YEAR ADJUSTED FOR HIDDEN ENVIRONMENTAL VALUES (G+H+I)		0.5
ADDENDA		
Social profit and loss account		649.0
M) Non-market benefits[3] net of compensations		
Recreation (388,655 visits @ 5,000 lire/visit)	971.5	
Mushroom picking without permits	50.5	
Protection ha160	160.0	
- Compensations	-533.0	
NET PROFIT/LOSS FOR THE YEAR ADJUSTED FOR NON-MARKET ENVIRONMENTAL BENEFITS (J+M)		649.5

Notes: 1. The annual risk depreciation charge was based on a 300ha 50-year old, evenly aged artificial spruce stand. Losses due to pestilence from similar stands amounted to 1,500ECU/ha, and the probability of a similar attack amounted to 1%. The charge is equal to 4,500 ECU (1,500ECU/ha x.0.01 x 300ha).

2. The non-marketable increase in growing stock was ascertained from forest inventories and a conservative valuation of 35,000 ECU was attached.

3. Non-market benefits were valued using the travel cost and contingent valuation methods.

Although important, attempts to report non-wood benefits by incorporating estimates of unpriced environmental impacts into an accounting framework, are not without problems. The single most important stumbling block to future development is measurement. Since at least the 1960s, accountants have been debating the issue of what measurement basis is most appropriate for accounting reports. In fact Houghton (1996) has observed that '*measurement has recently been called the most contentious issue facing the accounting profession*'. At present in Australia, an organization's balance sheet reports some assets at market exit prices, some at replacement cost, some at cash equivalent, some at present value, and other assets at historical cost. Consequently, it is argued that a balance sheet does not present stakeholders with any meaningful information (Chambers, 1996). Some accountants argue that adding the financial estimates of unpriced environmental benefits to balance sheet only exacerbates the current mathematical mismatch and should not be part of conventional accounting (Burritt and Cummings, 1999). In addition, accountants are traditionally conservative when preparing financial statements, a characteristic that most likely evolved to protect statement users against over-optimistic estimates of assets and revenues. Consequently, it will be a large undertaking to convince the accounting profession of the appropriateness of combining estimates of unpriced, non-wood benefits and financial information within a single reporting framework.

A reluctance to acknowledge the value of estimates of unpriced environmental impacts would appear to extend beyond the accounting profession. For example, estimates of unpriced environmental impacts have been used in extended cost-benefit analysis in several high profile resource use cases in Australia (e.g. mining at Coronation Hill, mining on Fraser Island). In each of these cases, the estimates of unpriced environmental impacts generated considerable debate and controversy as to their reliability and objectivity, and so in the end were not used to guide the resource use decisions that were made.

In acknowledgement of the difficulties in gaining acceptance of the reliability estimates of non-wood benefits derived, and the appropriateness of incorporating these values into traditional financial reports, the CRRP experiment and the Cansiglio forest experiment both used a reporting mechanism – a special purpose environmental report, and addenda respectively – that was an addition to the existing reporting framework. Until acceptance of alternative measurement techniques becomes more widespread, if it ever does, it would seem that attempts to report on non-wood benefits should run parallel to the existing reporting system, if they are to gain acceptance.

Another possible area for development is to review research being undertaken in other professions to determine if there are any other valuation techniques that are applicable to non-wood benefits, and might be more readily acceptable to accountants. This aim is not implausible because the accounting profession has a tradition of relying on other professions to contribute estimates of assets and liabilities to be reported in the balance sheet. For example, accountants rely on actuaries for the estimation of employee superannuation benefits, engineers for estimation of the value of specialised plant and equipment, and lawyers for estimating the value of contingent liabilities.

An additional consideration in using environmental economics valuation techniques to report on non-wood benefits is the suitability of the system for private sector and public sector organisations. Public sector forestry ventures usually operate with explicit multiple objectives (i.e. timber production, public recreation), they are funded with public money, their forests are accessible to the public, and so are highly visible, and they are a conduit for government policy. As a result, a clear case can be made for public sector forestry ventures to report on *all* aspects of their operations to interested stakeholder groups.

In contrast, the level of accountability of private sector forestry organizations (e.g. large public company, private unlisted company) is not clear. Proponents of environmental accounting would argue that the organisations are accountable to stakeholders in society, and so should report on the priced and the unpriced impacts of their operations. However, critics of environmental accounting would argue that private sector organisations' primary responsibility is to maximise the financial return earned by shareholders, and so any additional reporting is viewed as a tax. In order for an extended reporting system to work in a private sector organization, the chief executive officer and management would have to be committed to the concept of full communication with its stakeholders. Since it seems likely that the debate on the accountability of private sector organizations will remain unresolved, the best chance of successfully developing an environmental accounting system that incorporates non-wood benefits would appear to rest with public sector forestry organizations.

In conclusion, the problems with measurement and acceptance of environmental accounting systems that report financial estimates of non-wood benefits, need to be weighed against an increased expectation from stakeholders for accountability for all aspects of an organization's activities. Continual experimentation, such as that with the CRRP and the Cansiglio forest, will allow reporting systems to evolve that have greater acceptance by accountants, stakeholders and forest managers.

REFERENCES

Abt, C.C. and Associates (ACCA) (1972), *Annual Report and Social Audit*, Chicago: Abt.

Australian Accounting Research Foundation (AARF) (1997), *Self-generating and Regenerating Assets*, Exposure Draft No. 83, Melbourne: AARF.

Australian Accounting Standards Board/Public Sector Accounting Standards Board. (1998), *Accounting Standard AASB 1037, Self-generating and Regenerating Assets*, Melbourne: Butterworths.

Bebbington, J. and Gray, R. (1997), 'An account of sustainability: Failure, success and a reconceptualisation', http://les.man.ac.uk/IPA/papers/bebbin24.html, pp. 1—33.

Burritt, R.L. and Cummings, L.S. (1999), 'Environmental accounting, economic values and financial reporting – The Case of Earth Sanctuaries Ltd', paper presented at the 1999 Accounting Association of Australia and New Zealand (AAANZ) Conference,

Cairns.

Chambers, R.J. (1996), *Accounting, Evaluation and Economic Behaviour*, Houston USA: Scholars Book Co.

Creighton, C. and Sexton, G. (1996), 'Community rainforest reforestation program: A farm forestry demonstration', in *Managing and Growing Trees: Training Conference*, Volume I, Bundaberg, Australia, pp. 100—109.

CRRP Management Committee (1996), 'Community Rainforest Reforestation Program', *CRRP Annual Report 1995—1996*, Cairns Australia: North Queensland Afforestation Program Joint Board.

Deacon, R.T., Brookshire, D.S., Fisher, A.C., Kneese, A.V., Kolstad, C.D., Scrogin, D. , Smith, V.K., Ward M. and Wilen J. (1998), 'Research trends and opportunities in environmental and natural resource economics', *Environmental and Resource Economics*, **11**(3-4), pp. 383—397.

Dierkes, M and Preston, L.E. (1977), 'Corporate social accounting for the physical environment: A critical review and implementation proposal', *Accounting, Organisations and Society*, **2** (1), pp. 3—22.

Freeman, R. (1984), *Strategic Management: A Stakeholder Approach*, Marshall MA: Pitman.

Friedman, M. (1962), *Capitalism and Freedom*, Chicago: University of Chicago Press.

Godfrey, J., Hodgson, A., Holmes S. and Kam V. (1994), *Accounting Theory*, 2nd edn., Brisbane: Jacaranda Wiley.

Gray, R. (1994), 'Social and environmental accounting, accountability and reporting: New wine in old skins or silk purses from sows' ears', *Accounting Forum*, March: pp. 4—30.

Gray, R., Bebbington, J. and Walters, D. (1993), *Accounting for the Environment*, London: Paul Chapman.

Gray, R., Kouhy, R. and Lavers, S. (1995), 'Corporate social and environmental reporting; A review of the literature and a longitudinal study of UK disclosure', *Accounting, Auditing and Accountability*, **8**(2), pp. 47—77.

Gray, R., Owen, D.L. and Maunders, K.T. (1987), *Corporate Social Reporting: Accounting and Accountability*, Hemel Hempstead: Prentice-Hall.

Hanley, N., Wright, R.E. and Adamowicz, V. (1998), 'Using choice experiments to value the environment', Special Issue of *Environmental and Resource Economics*, **11** (3/4), pp. 413—428.

Harrison, S.R. (1999), 'Progress in estimation of intractable non-market values', in *Progress in economic science,* ed. S.B. Dahiya, Vol. 5, pp. 2719—2737, Rohtak: Spellbound.

Harrison, S.R., Herbohn, J.L., Emtage, N. and Smorfitt, D. (1998), 'Landholder attitudes to farm forestry and incentive schemes in North Queensland', in *Managing and Growing Trees Training Conference: Farm Forestry and Vegetation Management*, Kooralbyn Resort Hotel, 19—21 October.

Harrison, S.R., Miamo, J. and Anderson, M.W. (1999), 'Government and private sector joint venturing in natural resource development: The Queensland Plantation Forestry Joint Venture Scheme', *Economic Analysis and Policy*, 29(1), pp. 15—29.

Harte, G. and Owen, D. (1992), 'Current trends in the reporting of green issues in the annual reports of United Kingdom companies', in D. Owen (ed.), *Green Reporting*, Melbourne: Chapman and Hall, pp. 166— 200.

Herbohn, K.F., Harrison, S.R. and Herbohn, J.L. (2000), 'An alternative approach to accounting for natural resources: The case of multi-purpose forestry in Australia', *Society and Natural Resources,* 13(7), pp. 1—21.

Herbohn, K.F., Peterson, R. and Herbohn, J.L. (1998), 'Accounting for forestry assets: Current practice and future directions', *Australian Accounting Review*, 8(1), pp. 54–66.

Herbohn, K.F. and J.K. Herbohn (1998), 'Accounting for forestry assets: The development of an Australian policy', *Australian Forestry*, **61**(3), pp. 220–225.

Houghton, K.H. (1996), 'Measurement in accounting: A discussion of the theory and research evidence on measurement in accounting and financial reporting', Discussion Paper, The University of Melbourne.

Kirchloff, S., Colby, B.G. and LaFrance, J.T. (1997), 'Evaluating the performance of benefit transfer: An empirical inquiry', *Journal of Environmental Economics and Management*, 33, pp. 75–93.

Merlo, M. and Jobstl, H.A. (1999), 'Incorporating non-market values into the accounting systems of publicly and privately-owned forest enterprises: An operative stepwise approach', in C. Roper and A. Park (eds.), *The Living Forest: Non-Market Benefits of Forestry*, Proceedings of an International Symposium on Non-Market Benefits of Forestry, London: Stationery Office, pp. 341–372.

Milne, M. (1991), 'Accounting, environmental resource values, and non-market valuation techniques for environmental resources: A review', *Accounting, Auditing and Accountability*, **4**(3), pp. 81–109.

Milne, M. (1996), 'On sustainability; the environment and management accounting', *Management Accounting Review*, 7, pp. 135–161.

Palmer, J. (1998), 'The economics of environmental improvement: Will SMEs grasp the nettle of environmental accounting?', *Social and Environmental Accounting*, **18**(2), pp. 5–10.

Pearce, D. and Moran, D. (1994), *The Economic Value of Biodiversity*, London: Earthscan Publications.

QDPI (Queensland Department of Primary Industries) (2000), (http:// www.dpi.qld.gov.au/ forest/fid/plcychng.html).

Repetto, R., Magrath, W., Wells, M., Bear, C. and Rossini, F. (1989), *Wasting Assets, Natural Resources in National Income Accounts*, Washington: World Resources Institute.

Roberts, D.L., Staunton, J.J. and Hagan, L.L. (1995), *Accounting for Self-generating and Regenerating Assets*, AARF Discussion Paper No. 23, Melbourne: Shannon Press.

Roberts, J. (1991), 'The possibilities of accountability', *Accounting, Organisations and Society*, **16**(4), pp. 355–368.

Sterner, T. and van den Bergh, J.C. (eds.) (1998), 'Frontiers of environmental and resource economics: Testing the theories', Special Issue of *Environmental and Resource Economics*, **11** (3/4).

Ullman, A. (1985), 'Data in search of theory: A critical examination of the relationship between social performance, social disclosure and economic performance', *Academy of Management Review*, **10**(3), pp. 540–577.

DEVELOPING POLICIES TO ENCOURAGE SMALL-SCALE FORESTRY IN THE TROPICS

14. Taxation in the Forestry Setting

D.B. Smorfitt, M.L. Berry, A.J. Cummine and P.V. Townsend

In developing countries, decisions of landholders to establish forestry is in general not affected by taxation arrangements. However, in developed countries, taxation provisions can have strong and sometimes negative effects on private small-scale forestry, as illustrated with respect to an Australian case study. Taxation arrangements are found to be highly complex, and it is difficult to make generalisations about what deductions will be allowed or what tax on receipts will be payable; each case has to be viewed individually, and this increases the business risk in farm forestry. Anticipated changes hold the prospect of resolving two long-term disincentives, concerning use of the mechanism of *profit à prendre* for forestry rights, and opportunity for an active secondary market for immature plantations.

14.1 INTRODUCTION

Taxation legislation varies both within and between countries, making it a difficult topic to cover in a global context. Taxation issues play a more important role in developed countries, where there are many large commercial farm businesses, than in developing tropical countries where farming is frequently a subsistence activity. For many small-scale farmers – including most in developing countries in the tropics – taxation is not an issue in forestry decisions. Indeed, a high proportion of the timber grown typically is used on-farm, for which no income is reportable. On the other hand, in cases where a landholder faces a relatively high tax rate, the taxation regime can have a considerable impact on the attractiveness of forestry as an investment. A good example of this is the high rate of tree planting in the UK in the 1980s when tax deductions were allowed against very high marginal tax rates, as reported by Hill (2000).

Tax legislation is used in applying government policies to achieve political objectives such redistribution of wealth and economic development.

Tax legislation, irrespective of country of origin, has common characteristics such as income and revenue. Some differences relate to what legislation will allow as a deduction and what is regarded as income, and in the timing of the recognition of income and expenditure. This is of particular importance to the forestry sector due to the long lead-time from time of planting to harvesting, irrespective of whether planting for wood pulp or high-value timber products. Any attempts to legislate to defer deductibility of establishment costs until income is derived from the forestry activity will have a severe negative effect on revenue earned, especially for smaller farm forestry activities. Taxation can also be used as one of many means by which activities can be encouraged or discouraged. For example, allowing an increased level of deductions for plantation establishment and maintenance will tend to increase the rate of planting.

This chapter presents a case study of forestry taxation in tropical North Queensland, Australia, and highlights issues faced by taxpayers from a forestry perspective. Here, tax deductions for forestry expenditure and tax payable on timber sales are important considerations for forestry decisions. Tax impacts will depend on the role of forestry in the farm business. Landowners may be contemplating forestry as a means of diversifying, acquiring income or planning for retirement, or may have existing trees on their land that they wish to harvest. The disposal may be a one-off sale of timber, or ongoing transactions. Questions arise as to whether receipts will be taxed and whether expenditure will be allowed as a deduction.

This chapter deals with four areas of taxation in relation to forestry in Australia. Firstly, it deals with taxation of income derived from forestry operations and corresponding deductions. Secondly, it deals with activities that result in an inflow of funds from timber sales which, for various reasons, are regarded as a capital gain rather than income. Thirdly, current taxation issues together with future policy development and means likely to be adopted to overcome problems faced by the industry are examined. Finally the chapter deals with some recent developments in areas of taxation having implications for forestry.

14.2 INCOME, CAPITAL GAINS AND PRIMARY PRODUCTION IN AUSTRALIAN FORESTRY

In examining concepts of forestry income and deductions in Australia, it is necessary to consider sections of the *Income Tax Assessment Act 1997 (Cth)* that apply to income and deductions generally, specific provisions, case law and Australian Tax Office (ATO) Rulings and Determinations that relate to the forestry sector.

Income includes receipts such as the proceeds from rendering personal services, business receipts and receipts from use of property. Although capital gains are included in taxpayers' assessable income, it is still important to distinguish between income and capital, because capital gains attract beneficial tax treatment.

In Australia, classification as a primary producer results in advantages over and above other taxpayers. The benefits include the ability to base the rate of tax on a moving average of incomes to reduce the effect of yearly income fluctuations, and additional deductions allowable for primary production activities. In relation to forestry, under section 995 of the Act 'primary production business' includes carrying on a business of: planting or tending trees in a plantation or forest which are intended to be felled; or felling trees in a plantation or forest; or transporting logs for milling.

A taxpayer must demonstrate a direct involvement with the land, rather than simply be a transporter or manufacturer. Further, the taxpayer must be carrying on a business with these activities, c.f. planting trees for aesthetics or windbreaks, for example. Section 995 of the Act defines a 'business' as including any profession, trade, employment, vocation or calling, except being an employee. Taxation Ruling TR97/11 presents a number of factors that the courts have indicated are relevant in assessing whether a person is carrying on a business of primary production:

- whether the activity has a significant commercial purpose or character;
- whether the taxpayer has more than just an intention to engage in business;
- whether the taxpayer has a purpose of profit as well as a prospect of profit from the activity;
- whether there is repetition and regularity of the activity;
- whether the activity is of the same kind, and carried on in a similar manner, to that of the ordinary trade in that line of business;
- whether the business is planned, organised and carried on in a businesslike manner such that it is directed at making a profit;
- the size, scale and permanency of the activity; and
- whether the activity is better described as a hobby, a form of recreation or a sporting activity.

The courts consider these and other factors in determining whether a particular activity is a business. Presence or omission of any factor aids the court in arriving at a decision as to whether a business exists.

14.3 INCOME FROM THE SALE OF TIMBER FOR PRIMARY AND NON-PRIMARY PRODUCERS

The Australian Tax Office (ATO) may view income received from the sale of timber as either generation of income to be included in the taxpayer's assessment, or as a capital gain. Either way, the income or capital gain will be subject to taxation unless treated as an asset acquired before capital gains tax was introduced. If the taxpayer involved in forestry is to be assessed on income from the sale of timber, they would ideally want to be classified as a primary producer in order to gain the benefits associated with averaging of income and forestry specific deductions. However, income taxation may still be levied on taxpayers who are not regarded as primary producers, so both situations are discussed here.

Sources of Income for Primary Producers Involved in Forestry

For primary producers, Taxation Ruling TR95/6 identifies the following sources of income: proceeds from the sale of felled timber; proceeds from the sale of standing timber; royalties received from granting rights to other persons to fell; reafforestation incentive grants or payments; and insurance recoveries.

Proceeds from the sale of felled timber are to be included in assessable income in the year in which the timber is sold. The time of sale would generally be decided having regard to the general contract law, which is usually when property rights in the timber passes from one taxpayer to the next. A taxpayer involved in forestry operations may decide to sell the land and standing timber instead of felling the timber or selling the right to fell the timber. This transaction is not in the ordinary course of business, as the taxpayer is not in the business of trading land and timber. Standing timber is not considered to be trading stock. However, if the taxpayer meets the requirement of the trees having been planted and tended for the purpose of sale then the market value of the standing timber upon sale is to be included in the person's assessable income.

The common law concept of a royalty involves a payment made for the right to use a beneficial privilege or right. Importantly, the payment should be based on the extent of use of that right or privilege. Accordingly, payments not based precisely on usage would, strictly speaking, not be considered a royalty. However, most payments for felling trees are based on quantity or value of timber actually felled and would accordingly be assessed as royalty income. Reafforestation incentive grants or payments are also assessable as ordinary income.

Any insurance recoveries for loss of profits or income that would have been derived are assessable in the year of receipt. Section 70–115 specifically provides that insurance policy recoveries or any indemnity for the loss of trading stock that would have been included in the calculation of assessable income, must be included in assessable income. Thus a loss of stock of trees due to fire necessitates the subsequent insurance policy recovery to be included as assessable income. There is some relief in that, if the trees were destroyed by fire, the insurance monies received can be spread over a five-year period, to reduce the rate of tax.

Income from Timber Sales by Non-Primary Producers

Assessable income for a taxpayer who is not regarded as a primary producer may include proceeds from the sale of standing timber, profits from isolated transactions and royalties from granting rights to someone else to fell and remove timber (TR 95/6). Where a taxpayer purchases land in order to make a profit by selling the timber, the proceeds from the sale are assessable. Whether tax has to be paid on a one-off sale of timber is less clearcut. However, profits from a one-off sale may be assessable even if the taxpayer had no initial plans of sale, although profits from an isolated instance of selling standing timber may not be assessable if profit making by the sale of timber was never contemplated by the farmer. For example, if a taxpayer who is primarily involved in commercial cattle, decides to clear a block of land for pasture, the revenue from the sale of timber would not be assessable as ordinary income because profit making by sale was not contemplated at the time of original purchase of the land and trees.

14.4 INCOME DEDUCTIONS FOR FORESTRY

When assessable income is being determined, taxpayers may obtain both general deductions associated with carrying on a business and industry-specific deductions. In order to receive a *general business deduction*, it will have to be shown that either the outgoing and losses were incurred in producing the assessable income, or that they were necessarily incurred in carrying on a business for the purpose of gaining income. No deduction is allowable if the loss or outgoing is capital, private or domestic expenditure. Types of forestry expenses deductible include:

- preparation of site for planting, including deep ripping, mound ploughing, tilling and weed control. Although weed control and the other activities may take place a number of months prior to the actual planting of seedlings, it is part of the planting operations which must be distinguished from initial clearing;

- acquisition cost of seedlings;
- cost of planting seedlings;
- cost of fertilisers during the growth phase;
- cost of ongoing weed management, especially in the early years;
- pruning costs in later years;
- cost of thinning; and
- harvesting costs.

Expenditure on these items will be regarded as non-capital and, provided a link between the activities and the generation of income can be shown, would be deductible under s. 8-1. Some expenditures are, however, not deductible under this section due to being of a capital nature, e.g. cost of clearing the land of scrub in preparation for planting. Capital expenditure has been said to be expenditure relating to the establishment, replacement or enlargement of the business entity.

Specific deductions include depreciation, and expenditure on dams and irrigation channels, access roads, timber milling buildings, initial clearing and weed control, purchase of land and trees, and changes in trading stock.

Although expenditure relating to capital is not deductible under s. 8-1, the Act does provide some relief by allowing owners and quasi-owners to claim, as a deduction, depreciation of plant and articles, which would normally be capital expenditure. Essentially, depreciation is an annual deduction over the life of an income-producing asset as it is used up in the income-producing process (Woellner *et al.*, 2000). If an asset's effective life is less than three years, then 100% is deductible in the first year unless the taxpayer elects otherwise. Alternatively, the annual depreciation percentage varies from 10% to 60%, depending on effective life of the plant. There are however, some areas where primary producers are allowed additional write-off of expenditure, as discussed below.

Dams are generally regarded as being built principally for storage of water for the use in the business of primary production. Irrigation channels are regarded in a similar manner. Section 387-125 provides that expenditure on the construction of water facilities such as a dam is deductible in equal amounts over a three year period, starting in the year in which the expenditure was incurred.

A deduction is allowed for capital expenditure on forestry roads and for the construction or purchase of timber mill buildings, where the expenditure relates to the carrying on of timber operations to earn income. Forestry roads (including bridges, culverts or similar works forming part of the road) are roads constructed mainly to provide access to an area to enable either the planting or tending of trees or the removal from the area of felled timber. The road must be constructed primarily and principally for planting, tending, removing or milling felled timber. Timber mill buildings must be located in a forest where the trees are felled. The expenditure on forestry roads and timber mill buildings is deductible in equal instalments over the estimated period for which the access road and timber mill building was primarily built, or 25 years, whichever is less.

While expenditure on clearing and weed control prior to preparation for planting was previously deductible over a 10-year period, it is now considered to be of a capital or preliminary nature. Initial weed control as part of the clearing process must be distinguished from that associated with planting activities mentioned earlier in the discussion of general deductions.

Trading stock, both final (closing stock at year end) and initial (opening stock, beginning of the year) are used to calculate the cost of goods sold and thus the profit upon which the tax liability is based. The ATO regard only timber that has been felled as trading stock (TR 95(6)); standing timber is regarded as part of the land and as such is a capital asset and not trading stock. The felled timber is brought into account pursuant to s. 70-35. The excess of closing over initial is included as income in the taxpayer's assessment (s. 70-35(2)), while any decline is a deduction (s. 70-35 (3)).

Deductions are not normally allowed for expenditure incurred on the acquisition of a plantation or forest as this would be considered to be capital expenditure. However, provision exists in some circumstances for a deduction for the capital costs of acquiring land carrying trees.

14.6 CAPITAL GAINS TAXATION (CGT)

Capital Gains Taxation applies to any asset disposed of which was acquired after 19 September 1985. The manner of current disposal of pre-capital gains tax assets can have important tax implications for the taxpayer. For standing timber on freehold land, the owner may harvest the trees and sell them as sawlogs, or grant the harvest rights to someone else in return for a fee. The latter option is called a *profit à prendre* or 'a right to take something from another person's land'. The granting of the right creates the *profit à prendre*, which is an interest in the land. The creation of a *profit à prendre* may be viewed as both the acquisition and disposal of an asset, this being the *profit à prendre* itself, which may have capital gains tax (CGT) implications. As a result, there may be different taxation consequences, depending on the method of disposal of the trees.

Prior to the introduction of CGT, only receipts from the disposal of trees were considered to be taxable (s. 17).[1] Financial gains derived from the disposal of trees were often tax free (Mitchelson, 1996). In many instances, the gains from the disposal of timber or the granting of a *profit à prendre* to harvest timber would be a capital receipt rather than income. This was particularly true in instances where the landowner was a grazier, and the disposal consisted of a one-off transaction in order to clear land, rather than an attempt to make profit. In these instances, the receipt from disposal would be unlikely to be considered income and thus the taxpayer would be unlikely to incur any tax liability.

Capital gains taxation is complicated by whether the timber being disposed of is an asset in its own right, or part of the land, which may have been acquired prior to CGT. Even if the land was purchased prior to CGT, the trees may be treated as a post-CGT asset. The taxation implications are thus dependant upon the manner of disposal.

A capital gain occurs where the consideration on disposal of the asset exceeds the indexed cost base of the asset. The cost base of the asset is made up by the sum of the purchase price, incidental costs on both acquisition and disposal, capital expenditure incurred in enhancing the asset's value or in establishing or protecting the owner's title to the asset and, in some cases, non-deductible revenue expenses, including repairs and maintenance to non-income producing assets (Smorfitt and Berry, 1997). The landowner may also include expenditure on fertilisers, weed control, pruning and harvesting in the cost base of the trees, assuming these items are non-deductible expenses. This cost base, apart from the non-deductible revenue expenses, is then indexed to account for inflation.

Capital Gains Treatment of Disposal of Existing Trees

To facilitate the explanation of CGT consequences faced by landowners disposing of existing trees, the discussion is split into pre-CGT assets and post-CGT assets. Two scenarios may be considered for pre-CGT land and trees.

Sale of felled timber (CGT treatment). Suppose a landowner, who owns a pre-CGT acquired asset, land and trees, proceeds to fell the trees and sell the timber after 19 September 1985. The critical element to this scenario is whether the land and trees are seen as a single asset or as two separate assets. If the land and trees are seen as a single asset, does felling of the

---

[1] A multitude of commentators, such as Barkoczy and Bellamy (1994) and Slater (1993), have examined the vexed question of when proceeds from the sale of property are considered income, rather than capital. As Hill (1995) stated, the distinction between income and capital is unclear.

trees for timber split the asset in two and do create a new asset, leading to CGT liability. On the basis of ATO Tax Determination 93/79, a pre-CGT single asset of land and trees is regarded as two pre-CGT assets. Because there is no change in ownership due to the cutting of the trees, there is no disposal for CGT purposes and no CGT liability exists (s.160M). The subsequent sale of the timber is simply the sale of a pre-CGT asset and not subject to CGT (Smorfitt and Berry, 1997).

Granting of a profit à prendre. As an alternative manner of disposal, the taxpayer may decide to dispose of the timber via two contracts entered into post-CGT. One contract is for the sale of all the uncut timber, and the second grants the purchaser of the timber the right to enter the property over a period of time to remove the timber as required. It may be argued that this scenario is effectively the same as simply selling felled timber. Legally, however, it has quite different taxation consequences. Tax Determination 93/81 indicates that in this instance the two contracts are seen as one and, taken together, constitute a *profit à prendre*. Acquisition of the new asset (*profit à prendre*) by the grantor of the asset, occurs immediately prior to the time of the contract or, if not by contract, immediately before the time when the *profit à prendre* is vested in the grantee. This effectively means that the grantor of the *profit à prendre* disposes of a post-CGT asset and thus CGT applies. Taxation Determination TD 96/35 confirms this, and the fact that there are no further CGT implications for the grantor when the trees are actually felled. One important consequence of the creating of a *profit à prendre* by a landowner is that it creates an immediate tax liability. This liability will arise irrespective of whether the landowner actually receives any payment. This is because the CGT system is an accrual rather than a receipts system.

The CGT consequences of disposing of land and trees acquired after post-CGT also depend primarily upon the manner in which the disposal occurs, and again two cases may be considered.

Sale of felled timber. Land with trees acquired after 19 September 1985 represents a post capital gains tax asset. The proceeds of the disposal of this asset are subject to assessment. The felling of the trees is viewed as the splitting of the previous single asset into two separate assets. The subsequent sale of the timber and consequent CGT assessment provides for a portion of the original purchase price of the land and trees to be attributed to the trees. If the land is subsequently sold, the apportioned amount of the original purchase price of the land and trees (as a single asset) will be the original purchase price less that part of it attributed to the trees.

Granting a profit à prendre. Provided that the *profit à prendre* was granted after 20 September 1989, the treatment is the same as the earlier discussion, and thus the consequences of disposing of a pre-CGT asset by granting a *profit à prendre*, should be referred to.

14.6 THE IMPLICATIONS OF RECENT TAX REFORMS

In the year leading up to the final preparation of this chapter, Australia was immersed in a process of reform of the taxation system. The introduction of a goods and services tax to replace the myriad wholesale taxes and levies, an overhaul of the capital gains tax provisions, and the Ralph Review of Business Taxation (Ralph, 1999) have, taken together, created a lengthy period of uncertainty and disquiet for small businesses including primary producers.

Prior to the reform process, forest grower representatives lobbied the Federal Government to address issues associated with the tax treatment of forestry operations. Australian Forest Growers, the leading private forest grower association, had some notable successes including:

- persuading the Government in 1985 not to separate off-farm and on-farm income for tax purposes
- correcting highly adverse tax office interpretations of forestry operations in the drafts of forestry tax ruling 95/6
- securing the 1995 tax law amendment removing the anomaly that penalised the buying and reselling of standing plantations, and
- in concert with other associations and government agencies, successfully objecting to the destructive proposal in the 1999 Ralph Review of Business Taxation to defer forest establishment deductions until income is derived from the forest.

Legal anomalies or unintended and inequitable consequences of tax treatment of forestry operations still remained. These had been identified and nominated for attention in various federal or national reports and strategies including the report of the NPAC (1991), the Forest and Timber Inquiry conducted by the Resource Assessment Commission (Commonwealth of Australia 1992a), the National Forest Policy Statement (Commonwealth of Australia 1992b), the Wood and Paper Industry Strategy (Wood and Paper Industry Council 1995), and the national plantations strategy *Plantations for Australia: The 2020 Vision* (Common-wealth of Australia 1997). These issues included: the capital gains tax anomaly that penalises the use of *profit à prendre* as a means of granting a forest harvesting right; 'period inequity' still suffered by private forest operations with long periods between incurring costs and receiving income; 'lumpy returns' that cannot be offset by primary producer income averaging or the use of Farm Management Deposits; and the ongoing disincentive to trade in standing plantations.

The tax reform process not only put these reforms 'on hold' for many months, but in some cases, either eliminated the concern (or held promise of a solution when legislation is passed) or created new problems for forest growers. As at July 2000, the Federal Government was still sorting out the plethora of legislation to implement the GST and the first two stages of the government's response to the Ralph review; the government's stage 3 response to the Ralph review had yet to be announced, let alone legislated.

The policy and political environments are still remarkably fluid, and it is possible that some statements in the following sections could be out-of-date by the time of publication. A brief summary of the issues associated with changes currently being discussed and legislated as they relate to private forestry follows.

The Goods and Services Tax

The much publicised Goods and Services Tax (GST) introduced on 1 July 2000 is an important element in the new tax system. Except for more demanding and expensive requirements for keeping records and submitting tax returns, it appears that forestry is affected much the same as most other forms of primary production. Payments made by business operators generally include a GST component, or an input tax credit, whereas money collected from the sale of goods and services includes GST. Like other primary producers, forest growers will have to pay the GST on any goods and services they purchase to carry out their forestry operations, although the GST component can later be reclaimed as an input tax credit.

Pay-as-you-go tax collection

An integral part of the tax reforms is the new Pay-as-you-go (PAYG) tax collection system, meant to simplify payments to the Commonwealth and reduce the compliance costs of administering the taxes. Several payments are being brought under the PAYG system, including company tax instalments, provisional tax, instalments deducted from employees' wages, superannuation fund instalments, prescribed payments, reportable payments and any other withholding remittances. All of these tax details, plus the GST collected and the input tax credits of GST paid out in each accounting period, are combined to calculate the balancing figure in a single Business Activity Statement. This statement is prepared every month or quarter, depending on the size of the business, to determine the balancing payment paid or received from the ATO.

Registration for Australian Business Number and the GST
Business operators and primary producers having a turnover of more than $50,000 per year are required to register for GST and acquire an Australian Business Number (ABN), and registration for others is recommended. Below $50,000, registration is optional, but is necessary to recoup the GST paid on inputs.

Accrual versus cash accounting
The Government made a special provision under the new tax system whereby small business operators could choose either cash or accrual accounting procedures. The GST liability associated with each purchase and sale depends on which method the business operator chooses. Using cash accounting, the GST is to be paid in the period when money is received and paid. Using accrual accounting, the GST is due when an invoice is raised.

Positive and Negative Outcomes from the New Tax System

The existing primary producer exemption from Wholesale Sales Tax (WST) means that its removal and replacement with the GST after 1 July 2000 will confer little immediate direct benefit on primary production. The main benefits for forestry will flow from the removal of a number of State taxes over the next five years, from a reduction of 'embedded' taxes built into the costs of all industries, and from lower transport costs. The latter will result from the combination of an extension of the diesel fuel rebate scheme to rail transport, lower priced on-road fuel under the Diesel and Alternative Fuels Grants Scheme, for which primary producers registered for the GST will be eligible and, later, from the lower purchase prices for haulage vehicles following the removal of the WST.

For forest growers in particular, the new marginal income tax brackets could exacerbate the problems of period inequity associated with timber production. Townsend (2000) has modelled the impact of the new brackets and concluded that, overall, taxes on long-term forestry investments are likely to increase. The reason offered is that most harvest income will be taxed at the highest marginal rate, while deductions for establishment and maintenance, if claimed on incomes less than $60,000, will be less in future because the marginal rates below this threshold have been reduced.

Impacts of the New Business Tax System on Profitability of Farm Forestry

The Review of Business Taxation, conducted by the Ralph Committee (Ralph, 1999) recommended the implementation of a number of structural changes to the business tax system. One of the major changes introduced so far is the replacement of accelerated capital depreciation allowances with a phased reduction in the company tax rate to 30% by 1 July 2001. Small businesses will also have access to a number of measures under the Simplified Tax System (STS). Changes in the method of asset depreciation allow items costing $1,000 or less to be written off immediately. Other assets with effective lives of less than 25 years can be included in a pool and depreciated at a rate of 30% declining value method.

A number of other specific measures are expected to affect the forest and timber industries. They include: the 'integrity measures' legislation; new legislation for *profit à prendre* and the treatment of rights; review of the tax provisions applying to the purchase and resale of immature plantations; the tax treatment of co-operatives; new arrangements for capital gains that benefit long-term investments; a pooled development fund structure to attract venture capital from pension funds; a major project facilitation process to replace accelerated depreciation allowances for capital; and alignment of foreign investment rules with those of Australia's major competitors. The first three of these are discussed below.

'Integrity measures' legislation

Of particular and specific importance for forestry operations are the 'integrity measures' designed to minimise 'leakage' from public revenue, under the *New Business Tax System (Integrity Measures) Bill 2000*. Schedule 1 of the Bill deals with a new tax treatment of so-called non-commercial losses, while Schedule 2 eliminates the '13-month' rule for the deduction of prepaid expenditure. Loss of the 13-month rule created concern in the tax-effective investment sector, of which the prospectus-financed plantation companies are a substantial element. Despite this, the afforestation investment companies immediately began the process of restructuring their business operations and their marketing programmes such that, from 1 July 2000, afforestation prospectuses will be offered to potential investors throughout rather than at the end of a financial year.

Although designed to cover all potentially 'non-commercial' business activity (including artists, musicians, part-time sales agents, and many others), Schedule 1 has major implications for small-scale primary producers with off-farm income. The new legislation is likely to enhance the potential of farm forestry, with a long lead-time between expenditure and income, as a possible solution to the problem many small landholders will now face of how to remain primary producers in the face of the ill-conceived threshold tests.

Treatment of rights

The tax treatment of *profit à prendre* has been a source of difficulty to the forestry sector since 1992, when it was determined that the grantor of a *profit à prendre* was deemed to have disposed of a new capital asset and thus to have incurred a tax liability, irrespective of whether any payment had been received at that time. A further anomaly is that, in addition to the up-front capital gains tax liability, the grantor is technically liable for income tax on the payment actually received when the trees are felled. The Review of Business Taxation recommended a change to the taxation treatment of rights, which in forestry include not only *profit à prendre* but also carbon rights, rights for environmental services such as salinity amelioration and biodiversity, and agreements to grow trees in return for an annuity plus some royalty payment. The Ralph review recommended that the potential capital gains tax liability should be removed and that income should be taxed across the period of the right (up to a period of 10 years). At the time of writing this chapter, the Government was still considering how to legislate to implement this recommendation. Full implementation of the recommendation could be a beneficial outcome for forestry in general and for small-scale forestry in particular.

Trading in immature plantations

The legal anomaly that had technically penalised the purchase and resale of immature plantations was removed in 1995. Nevertheless, the tax treatment of such transactions remained inequitable in relation to the establishment of new plantations. Purchase and sale of immature plantations are currently treated under the trading stock rules, thus requiring the purchase price to be carried forward until the plantation is resold or harvested. This contrasts with the immediate deduction available to a taxpayer when establishing a new plantation. This inequity has long been offered as the reason for there being no 'secondary market' for immature plantations, the lack of such trade discouraging investment in new long-rotation plantations compared with short-rotation pulpwood tree crops. The opportunity exists in the Government's Stage 3 response to the Ralph review to address this inequity, by bringing forward the tax deductibility of the purchase price or recognizing traded immature plantations as capital assets rather than as trading stock (Townsend, 2000). If this opportunity is taken to stimulate the much-needed creation of a secondary market in plantations, it would in turn make investment in long-rotation high-value farm plantations more attractive.

14.7 CONCLUDING REMARKS

For many small-scale farmers in the tropics, taxation arrangements do not enter into their forestry decisions. However, in some cases, such as farm forestry in North Queensland, Australia, taxation impacts can be large and can be a disincentive to forestry investments.

Recent years have seen positive change in the tax treatment of private forestry in Australia. It may take a number of years for the benefits of recent changes to be realised. Further change is anticipated as part of the Federal Government's Stage 3 response to the Review of Business Taxation. These anticipated changes hold the prospect of resolving at least two long-term problems - the disincentive to use the mechanism of *profit à prendre* for forestry rights, and the disincentive to participate in an active secondary market for immature plantations. Hence they should make investment in private forestry more attractive, for example, through new arrangements for capital gains that benefit long-term investments, and through a pooled development fund structure to attract venture capital from pension funds. The problem of 'uneven returns' will continue to be an irritant for private forest growers, making forest planting and management phased over a number of years appear preferable from a tax perspective to the one-off 'superannuation' farm plantations used in the past.

Regardless of how growers approach their farm forestry enterprise, the standard warning that the tax treatment of any private forestry enterprise will be unique to that individual enterprise, should always be heeded. Parameters such as scale of operation, establishment and management costs, stand management regime, income from other farm activities, level of off-farm income, accounting method used and levels of equity and borrowing will have to be considered. Often, the tax treatment can only be resolved once a tax return has been submitted and processed. The uncertainty about tax liability can add to investment risk in forestry.

REFERENCES

Barkoczy, S. and Bellamy, N. (1994), The Post Myer Trilogy:Justice Hill's Interpretation of The Myer Emporium Ltd v. GCT. *Australian Tax Review,* **23**(1), pp. 7–21.

CCH Tax Editors, *Australian Master Tax Guide 1997*, CCH Australia Limited, Sydney.

Commonwealth of Australia (1992a) *Forest and Timber Inquiry*, Resources Assessment Commission, Canberra: Australian Government Printing Service.

Commonwealth of Australia (1992b) *National Forest Policy Statement· a New Focus for Australia's Forests*, Canberra: Australian Government Printing Service.

Commonwealth of Australia (1997), *Plantations for Australia – the 2020 vision,*

Canberra: Department of Primary Industries and Energy.

Hill, D.G. (1995), 'Income and Capital: Have the Goal Posts Been Moved?', *Taxation in Australia*, **30**(3), pp. 152–153.

Hill, G.P. (2000), 'Policies for small-scale forestry in the United Kingdom', in S.R. Harrison, J.L. Herbohn and K.F. Herbohn, (eds.), *Sustainable Small-scale Forestry: Socio-economic Analysis and Policy*, Edward Elgar, Cheltenham.

Mitchelson, P. (1996), Income Tax Law for Accountants, John Wiley & Sons, Brisbane.

NPAC (National Plantations Advisory Committee) (1991), *Integrating forestry and farming: commercial wood production on cleared agricultural land*, Main report, Appendix A, Appendix B and Appendix C, Canberra: National Plantations Advisory Committee, Department of Primary Industry and Energy, Commonwealth of Australia.

Ralph, J.T. (1999) *A tax system redesigned: more certain, equitable and durable; report / Review of Business Taxation*, Canberra: Australian Government Printing Service, 4 vols., http://www.rbt.treasury.gov.au.

Slater, A.H. (1993), 'A Spreading Stain: The Character of Income', *Australian Tax Review*, **22**(2), pp. 132–159.

Smorfitt, D. B. and Berry, M. L. (1997), A review of the capital gains tax treatment of the disposal of existing trees by landholders. *Australian Forestry*, **60** (4), pp. 213–217.

Townsend, P.V. (2000), 'A Tax System Redesigned: Implications for Australia's Forest Industries', *Opportunities for the New Millennium*, Australian Forest Growers Biennial Conference, Cairns, 4 – 6 September, 2000.

Woellner, R., Barkoczy, S. and Murphy, S. (2000), Australian Tax Law, 10th Edition, Sydney: CCH Australia Limited.

Wood and Paper Industry Council (1995), *Wood and Paper Industry Strategy*, Canberra: Ministry for Industry, Science and Technology.

15. Landholder Attitudes to and Participation in Farm Forestry Activities in Sub-Tropical and Tropical Eastern Australia

N.F. Emtage, S.R. Harrison and J.L. Herbohn

Australian governments at all levels are promoting farm forestry for a number of economic, social and environmental objectives. However, the rate of establishment of plantations on private land in the tropics and sub-tropics remains low. A number of studies have investigated factors affecting adoption of farm forestry by landholders in these regions with the aim of facilitating greater participation. These studies indicate that most of plantation establishment on private land is by specific types of landholders. The majority of landholders are not convinced that plantation development is financially viable and are content with their current lifestyle. Some groups believe that forestry is not a suitable activity for agricultural land.

15.1 INTRODUCTION

Australian federal, state and local governments are promoting farm forestry in an effort to achieve a number of objectives in relation to environmental protection, biodiversity conservation, regional development and reduction in timber imports (Commonwealth of Australia, 1992a,b, 1997). Private land that has been inappropriately cleared in the past is the focus of the government efforts to promote the private plantation estate. Australian governments see the expansion of farm forestry as a way of achieving a number of these public objectives simultaneously. The problem facing governments is that private landholders in highly productive sub-tropical and tropical regions of eastern Australia have been less than enthusiastic about plantation development on their land despite active promotion of and assistance for farm forestry by State government land and forest management agencies. Landholder lack of response to many of these promotions has convinced many researchers of the need to understand better landholders' visions of the appropriate role for farm forestry on private lands, including their motivations for planting, perceived impediments to planting, and the types of incentives they desire. The situation is complicated by the variations in attitudes and socio-economic circumstances within the rural community. Some landholders are enthusiastic about farm forestry while others are firmly opposed to plantation development on private agricultural land.

Up until the 1960s, there had been major land clearing programmes in Australia, some funded by government, to make land available for agricultural expansion. The traditional landholder ethic has been to carry out 'timber treatment' and 'regrowth control'. Progressively since the 1960s environmental protection and conservation have established as norms in many parts of Australian largely urban-based society.

While the population of landholders in the higher rainfall areas near the coast of Queensland and northern New South Wales includes traditional farmers, there has been considerable land subdivision and population growth, which has involved an influx of new landholders. The demographic mix of these areas has changed with the majority of newcomers being from the major cities of Sydney and Melbourne. Many city people are attracted to these areas which they envisage will be cleaner and healthier places to raise their families, less affected by pollution and overcrowding problems of the big cities, and the social problems of juvenile delinquency and drug use. They bring with them an idealised picture of country lifestyles. Living in the country is viewed as providing an opportunity to be 'closer to nature' as well as operate small agricultural operations. Many continue to work off-farm in regional centres. Lower reliance on property income together with urban backgrounds leads to differences in value systems and different objectives for land management compared to 'traditional' landholders.

The above factors have contributed to strong interest in growing trees on farms. It is generally agreed that such activity is socially desirable in that it generates positive externalities. It would seem that with a little assistance from government – as subsidies, extension activity or creating a facilitating environment – landholders would have a rapid uptake of farm forestry. In practice, the rate of tree planting appears to have been somewhat less than expected, hence there is a need to examine what impediments exist, in terms of landholder attitudes to and resources for farm forestry. This chapter is concerned with the former of these impediment types.

A number of researchers in the field of rural extension have described the need for forestry support and extension programs to be targeted to the specific requirements of landholders (Byron and Boutland, 1987; Chamala, 1987; Stewart and Reid, 1994; Kaine and Lee, 1995; Vanclay, 1997;

Vanclay *et al.*, 1998). A prerequisite for this targeting is a thorough understanding of landholder attitudes to farm forestry.

This chapter reviews the findings of studies of landholder attitudes to farm forestry and to government support programs, and the uptake of farm forestry, in tropical and sub-tropical eastern Australia. Efforts to identify groups of farmers with relatively homogeneous attitudes to farm forestry, which may assist in targeting assistance programs and improving understanding of the socioeconomic factors affecting farm forestry practice, are also examined.

15.2 STUDIES OF LANDHOLDER ATTITUDES TO FARM FORESTRY AND OF FARM FORESTRY PRACTICES

A number of recent studies in Australia have specifically investigated landholder attitudes to farm forestry in tropical and sub-tropical eastern Australia (Table 15.1). A range of techniques have been used in these studies including group interviews, focus groups, telephone surveys, mail surveys and combinations of these.

Table 15.1 Studies of landholder attitudes to farm forestry in tropical and sub-tropical eastern Australia

Researchers and year	Study location	Sample size
Harrison, Sharma and Lamb (1992–93)	Obi Obi catchment (south east Qld.)	143
Broome (1993)	Atherton, Eacham and Johnstone shires (far north Qld. region)	99
Eono (1995)	CRRP shire mayors	10
Emtage (1995)	Richmond River catchment landholders (north east NSW)	102
ABARE (Wilson and Yannick) (1995)	All regions in Australia	2,000
Eono (1996)	CRRP shires	10
Specht and Emtage (1998)	Northern rivers region landholders (north east NSW)	186
Emtage, Herbohn, Harrison and Smorfitt (in prep.)	Atherton, Eacham and Johnstone landholders (far north Qld. region)	224

The national survey of Wilson and Yannick (1995) reported substantial variation in farm forestry participation rates across Australia. Over 50% of landholders are active farm foresters in some regions of Victoria, southern New South Wales and Western Australia, yet less than 10% of landholders practice farm forestry in northern New South Wales and Queensland.

Landholder Reasons for Planting Trees

It has been generally observed that environmental benefits are perceived by landholders as the most important benefit of tree planting. For example, Harrison *et al.* (1994) found that Obi Obi Valley landholders consider the most important reasons for tree planting are to protect and restore land, to attract wildlife and birds, personal interest in trees, watershed protection and to improve the look of the property. Economic motives such as to make money and to diversify the farm business are ranked much lower (Harrison *et al.* 1994, Harrison and Sharma, 1995). Similarly, in a survey of landholders participating in the Community Rainforest Reforestation Program (CRRP) in three North Queensland shires by Broome (1993), land and stream protection and the provision of shade and windbreaks were viewed as the most important benefits of tree planting.

These planting motivations are common Australia-wide, although it appears that the provision of shelter for livestock is perceived as being more important by landholders away from sub-tropical and tropical regions (Wilson and Yannick, 1995). A pilot study by Eono and Harrison (1996b) found that providing a habitat for wildlife and protecting soil and water quality were perceived by landholders as being the most important private benefits of tree planting in North Queensland. Specht and Emtage (1998) reported similar results for tree planting in northern NSW with the highest rated reasons being soil and land protection, creation of wildlife habitat and farm beautification. Emtage *et al.* (in prep.) also found that landholders rate 'environmental' and 'personal' reasons as the most important reasons for planting trees in North Queensland.

Economic benefits such as the creation of a timber resource and income generation are rarely mentioned as benefits of tree planting. In most of the studies reviewed the lowest rated reasons for planting are for various types of commercial timber production. There appears to be a number of reasons for the lack of importance placed on planting trees for commercial reasons by landholders in these surveys. Part of the explanation is landholder perceptions of critical restrictions to plantation development such as the lack of access to markets and perceived sovereign risk. Other reasons include the attitude of some groups within the community that agricultural land should be used exclusively for agriculture and landholder perceptions that they have sufficient tree cover to meet their requirements.

A consequence of reluctance of landholders to plant trees for commercial purposes is that the number of trees being established on private land in tropical and sub-tropical areas of Australia is low. Specht and Emtage (1998) investigated the number of trees being planted for various purposes in the Northern Rivers region of New South Wales. They reported that although tree planting for commercial purposes is rated as relatively unimportant by most landholders, the number of trees established for timber production and other commercial enterprises (such as fruit, nut and oil production) is far greater than the number of trees established for other reasons (Table 15.2). The figures are skewed by the size of several large plantations, accounting in aggregate for over 30,000 trees. It is apparent from these data that unless landholders are planting trees for commercial purposes the number of trees established (and area planted) is unlikely to be sufficient to meet the government objectives described earlier.

Table 15.2 Total and average number of trees planted for various purposes by private landholders, Northern Rivers region, NSW

Purpose of planting	Total number of trees	Average number of trees per farm
All timber only	167,700	2,800
Commercial non-timber	69,400	1,980
Stock and crop shelter	16,700	220
Habitat and regeneration	3,900	100
Property beautification	3,850	50
Beautification and other reasons	2,900	70
Timber and other reasons	2,440	240
Other	2,400	235
Own fruit or nuts	1,900	30
Soil protection	1,300	60
Stock/crop shelter and other reasons	50	25

Source: Specht and Emtage (1998).

Perceived Restrictions to Tree Planting on Private Land

A number of surveys have examined landholder perceptions of constraints to tree planting (including Harrison and Sharma, 1995; Emtage 1995; Wilson and Yannick, 1995; Specht and Emtage, 1998; Emtage *et al.*, in prep.). Perceived constraints to private plantation development include financial concerns (including concerns about harvest rights and availability of markets), satisfaction with current land-use activities and lifestyles, and for some the perception that forestry is not a legitimate activity for agriculturally productive land. The importance of financial concerns has been noted by Harrison *et al.* (1994), Emtage (1995), Wilson and Yannick (1995), Specht and Emtage (1998) and Emtage *et al.* (in prep). Surveys reveal that farmers are worried about time taken for trees to grow to harvest age, during which they not only have to finance establishment of the trees and cover the costs maintaining the plantation, but also carry these costs for about 25 to 35 years, with perhaps a little earlier thinnings revenue.

Many landholders are also concerned about the markets for timber products. Public agencies have long dominated the Australian timber industry and have set low log prices. Landholders on larger properties often have some areas of native forest that they have logged in the past, from which returns have been discouragingly low. These concerns are increased by the experiences of landholders that established plantations with *Pinus* species in the 1960s and 1970s and now find they cannot sell their timber.

Related to these financial concerns is uncertainty about the regulations pertaining to management of freehold land and in particular concern that changes to land management legislation or regulations in the future will prevent the harvesting of trees. The New South Wales government has enacted harvest guarantee legislation in an attempt to alleviate these concerns but many landholders remain suspicious about the likely court interpretation of this legislation and about future watering down of harvest rights (Emtage and Specht, 1998). State government regulation of land clearing and forestry operations on private and public land is highly contentious for rural communities and uncertainty about these regulations raises concerns that there is a high sovereign risk associated with investment in forestry.

The dominance in forestry of State forestry agencies in Australia and the extent of state and local government land management regulations has led to the perception in sections of the rural community that forestry is an activity only appropriate for governments and corporations. The latest national and state government policies to establish plantations on previously cleared agricultural land are resented by many in rural communities. Many rural people believe agricultural land should only be used for agriculture and forestry should be restricted to 'unproductive' land (Emtage and Specht, 1998). With the financial returns for many agricultural activities low through the 1990s, the price of rural land dropped in many regions of northern New South Wales and south-east Queensland. A number of landholders are disturbed by forestry agencies and private companies buying farms to establish plantations. Some resent the loss of previous neighbours and the decrease in agricultural activity in these areas (Emtage and Specht, 1998). In times of economic hardship the loss of neighbours increases landholders fears that they may be next to be forced from the land.

As well as the perceived social disruption, there is concern that rural communities will suffer financially from plantation development. There are fears that shops and services will be lost because of a fall in rural population, and local government rate revenues will decrease due to rate exemptions for State forest agencies. In addition some landholders and local councils believe that log transport in the future will damage rural road systems (Emtage and Specht, 1998).

Doubts about financial returns are exacerbated by the failure in many cases of forestry agencies and companies to offer landholders appropriate production technology. With the financial returns from timber products typically taking 25–35 years for sawlogs, landholders are looking for earlier non-wood plantation benefits, such as shelter benefits for crops and livestock, and improvement of the aesthetic and environmental values of their properties. There have been some publicly funded tree planting schemes that have explicitly sought to improve the conservation values of landholdings, such as the Community Rainforest Reforestation Program and the Wet Tropics Tree Planting Scheme in North Queensland. Most schemes however, including joint venture programmes of state forest agencies, are perceived by landholders to have low environmental and aesthetic values and not be designed to integrate with other farm activities (Specht and Emtage, 1998).

Many landholders believe that the appropriate place for farm forestry is in areas of their properties that are not suitable for agriculture because they are either too steep or have low soil fertility (Emtage and Specht, 1998). Many of these areas on private land are prescribed as 'protected' or environmentally sensitive by state land management agencies and forestry operations on them are highly restricted or banned. Government regulation of activities on freehold land in eastern Australia has expanded rapidly since the late 1980s. Forestry and land clearing are not the only activities that are now more tightly regulated. New restrictions have been placed on property rights with respect to landholders freedom to access and dam water, their burning practices, their use of pesticides and many other farming activities. These new restrictions have made landholders feel their knowledge and land management skills are being rejected or ignored. The regulations increase resentment that the 'bush' is being controlled by city-based values that are not relevant to rural communities and not informed by direct experience. The gap between landholder visions of appropriate farm forestry development and that of the government and urban-based community is another factor that diminishes landholder confidence in plantation development.

15.3 RELATIONSHIPS BETWEEN SOCIO-ECONOMIC CHARACTERISTICS AND FARM FORESTRY ATTITUDES AND BEHAVIOUR

Despite the problems described above, there are a small number of enthusiastic farm foresters in the rural community in sub-tropical and tropical eastern Australia. Examination of the socio-economic characteristics of these landholders can help those planning and administering farm forestry policies and development programmes to understand the likely impacts of policy alternatives on landholders and to understand both the current adoption of farm forestry and means to promote the activity.

Some previous studies have examined the relationships between socio-economic characteristics and farm forestry attitudes and behaviour. Wilson and Yannick (1995) examined the differences between those who had planted more than 20 trees in the 1993–94 financial year and those who had not. Those planting trees were found to:

- have smaller landholdings;
- be younger;
- have more valuable land and make more money per hectare;
- use a higher proportion of land for cropping; and
- have greater sheep and lower cattle numbers.

Other studies support the finding that those who report planting trees typically have smaller properties (Harrison *et al.,* 1994; Emtage, 1995; Specht and Emtage, 1998; Emtage *et al.*, in prep.). Also, studies have identified relationships between tree planting behaviour and other socio-economic characteristics (Dunchue, 1990; Stewart and Reid, 1994; Dunchue and Sinclair, 1995; Specht and Emtage, 1998; Emtage *et al.*, in prep). Those who have planted trees tend to:

- use more 'outside' information to assist land management decisions;
- earn a higher proportion of income off-farm;
- have attained a higher than average level of formal education;
- be members of more than the average number of land management groups; and
- anticipate family involvement in future land management.

Differences in attitudes between those who have and have not planted trees on their properties have also been examined in a number of studies (Emtage, 1995; Specht and Emtage, 1998; Emtage *et al.*, in prep.). These studies have revealed that landholders that are active tree planters attach higher importance on planting for 'environmental' and 'personal' reasons.

Further, they are less concerned about economic constraints on planting in general, less satisfied with their current agricultural activities, and less concerned about the potential loss of flexibility in future land-use.

15.4 VARIATIONS IN RURAL COMMUNITY ATTITUDES TO FARM FORESTRY

Grouping landholders according to their attitudes to farm forestry can help decision-makers understand the present state of development of farm forestry. Those planning and administering publicly funded farm forestry development programmes can understand better the socio-economic impacts of various types of incentive programmes on different sections of the rural community. Programmes can be designed which meet the needs of particular types of landholders and appropriate communication strategies can be developed to promote them.

Analysis of the socio-economic characteristics and attitudes of those currently involved in the development of plantations on private land indicates that particular types of landholders are establishing the majority of plantations. In an effort to understand better the variations in attitudes and behaviour in regards to farm forestry, Emtage and others have carried out a series of studies to identify and describe distinct groups within the community with regard to attitudes to farm forestry and farm forestry behaviour. The basis for defining these groups is cluster analysis of respondents' ratings of the importance of tree planting for various reasons (used by Emtage, 1995; Specht and Emtage, 1998), and for Emtage *et al.* (in prep.) also their rating of the importance of various potential constraints to planting. Cluster analysis can be used to group respondents with similar responses to the topics used as the basis for their definition (i.e. ratings of the importance of various reasons for and restrictions to plantation development). These studies have identified five landholder types or groups which are relatively homogenous internally but vary between groups in terms of attitudes to farm forestry. Figures 15.1 and 15.2 (from Emtage *et al.*, in prep.) illustrate the variation in importance of various reasons for and constraints to planting found for these groups in North Queensland.

In Figure 15.2, 'Satisfied/flexibility' refers to satisfaction with present lifestyle and a desire to maintain flexibility for future land use decisions. 'Lack labour, finance, equipment' refers to a lack of these resources to establish plantations. 'Fire/pest risks' refers to the perception that these potential problems could affect the success of plantations.

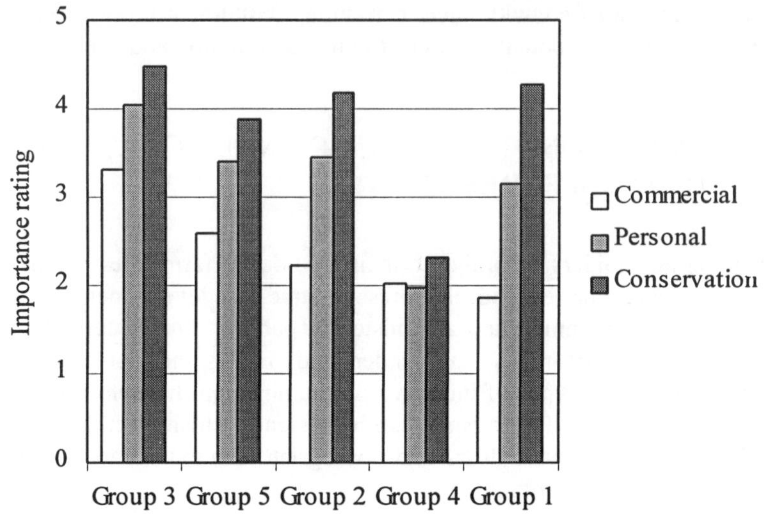

Figure 15.1 Average importance of various reasons for planting by groups of landholders defined through cluster analysis

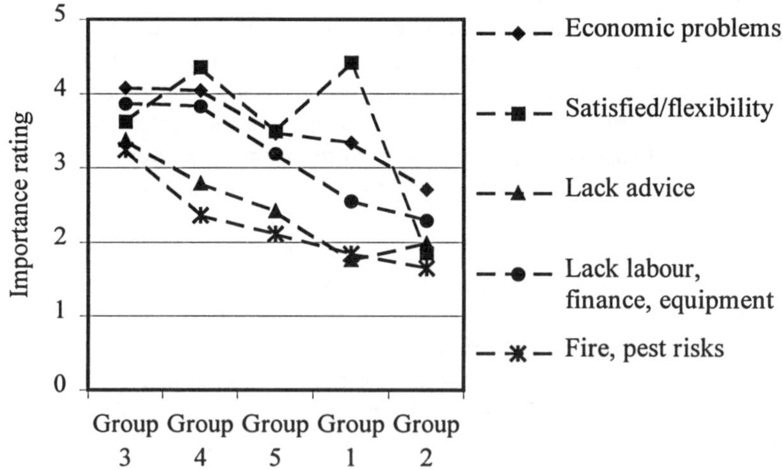

Figure 15.2 Average ratings of importance of various restrictions to plantation development, North Queensland landholder groups

The five groups all rank the different scales for planting trees in the same order except group 4, who rated the importance of planting for 'personal' reasons slightly lower than the importance of planting for 'commercial' reasons. 'Conservation' is the highest rated scale for planting for all groups, and the ratings for the 'commercial' scale are the lowest for all but group 4 (Figure 15.1). Variation between the groups is mainly due the degree of importance attached to planting for commercial reasons. Here group 3 clearly exceeds the other groups. Group 3 also indicates the greatest interest in planting trees for personal reasons.

While information problems, satisfaction with current activities and the potential loss of flexibility and economic problems are equally important restrictions to planting for respondents as a whole, there is clear variation in the rating of importance given to these potential restrictions by the various groups (Figure 15.2). Again group 3 stands out in particular, attaching greater relative importance to all potential restrictions to planting except the scale relating to less satisfaction with current activities and concern about loss of flexibility for future land management decisions. Concern about these topics is of considerable importance to members of groups 4 and 1. Group 2 had the lowest rating of importance for all scales relating to restrictions to tree planting.

Having defined groups of respondents on the basis of their attitudes to farm forestry (ie. ratings of importance for various reasons for and restrictions to plantation development), the groups were then assessed to determine if they differed in terms of their socio-economic characteristics. Significant differences were found in the socio-economic characteristics of the various groups. Tables 15.3 and 15.4 (from Emtage *et al.*, in prep.) report some of the defining characteristics of these groups. The groups differ in terms of the level of tree planting they have undertaken (Table 15.4) and in other socio-economic characteristics. At present most planting of commercial woodlots is being undertaken by landholders that could be termed 'hobby' or 'part-time' farmers in group 2 (Tables 15.3 and 15.5). This group of landholders appears to have higher off-farm income and hence the financial resources to undertake planting as well as attitudes compatible with farm forestry development (Table 15.3). This is reflected by the fact that this group also maintains the highest proportion of their landholding under native forest (Table 15.3). A few members of other groups are also planting commercial timber lots (Table 15.4), but their forestry activity appears to be restricted by a combination of economic factors, their time of life (i.e. old age), uncertainty about future regulations and markets, and their desire to maintain their present lifestyles.

Table 15.3 Selected characteristics of North Queensland landholder groups identified through the use of cluster analysis

Group number	Land size (ha)	Cropping (% of holding)	Native forest (% of holding)	Time managed (years)	Income (% from holding)	Family work hours per week
1	58	47	9	16.5	45	60.1
2	54	16	31	14.4	36	44.8
3	81	37	11	18.4	54	64.4
4	100	45	25	27.0	71	99.1
5	74	38	27	21.5	62	54.5
Mean	69.2	35	25	19.3	53	56.9

Table 15.4 Proportion of respondents who have planted more than 30 trees and future planting intentions by landholder groups

Group number	Proportion planted more than 30 trees (%)	If intend to plant for timber and other reasons (%)	If intend to plant for conservation and other reasons (%)
1	34	19	22
2	55	30	45
3	45	25	38
4	21	7	0
5	29	10	14

As part of the study of landholder attitudes to farm forestry in North Queensland (Emtage *et al.*, in prep.), locally based farm forestry extension personnel were asked to help name the five groups identified through cluster analysis of survey responses. The names arrived at by the extension officers, and the main influences identified for planting behaviour, are presented in Table 15.5.

15.5 INCENTIVES TO PROMOTE FARM FORESTRY DEVELOPMENT

An understanding of landholder attitudes to government incentives for tree planting can assist in the design of cost-effective intervention programs. Harrison *et al.* (1994), Emtage (1995), Specht and Emtage (1998) and Emtage *et al.* (in prep.) have used surveys to elicit landholder attitudes to a number of potential tree planting incentives. The responses to the surveys have been surprisingly consistent. It would appear that greater harvest security and economic incentives such as tax breaks and higher prices for timber products would be of greatest assistance in developing farm forestry (Harrison *et al.*, 1994, Specht and Emtage, 1998, Emtage *et al.*, in prep.). Joint venture arrangements are consistently the lowest rated incentives in these studies. Landholders appear to place high priority on retaining control over the management of their property including the species and area planted, planting design and timing of harvests, and do not wish to share income from harvests. Many of the silvicultural decisions would be outside the landholders' control in joint venture arrangements.

Table 15.5 Landholder group names given by extension personnel and main influences on planting behaviour

Group	Influences on planting behaviour
1 High intensity farmers	Have some personal interest in trees but also a strong commercial focus. Limited capital and property size leads to risk aversion. Enjoy agricultural pursuits.
2 Retired professionals and hobby farmers	Strong personal interest in tree growing and lower reliance on landholding for income leads to high participation in farm forestry.
3 Progressive second generation farmers	Strong interest in tree growing but have high reliance on landholding for income and low ability to cope with demands of planting and management.
4 Traditional	Low personal interest in tree growing. High reliance on land for income. Enjoy agricultural production. Interest in managing native forests.
5 Experienced/ comfortable	Moderate personal interest in tree growing and reliance on land for income. Low labour input relative to income produced.

Specht and Emtage (1998) and Emtage *et al.* (in prep.) recommended packages of incentives tailored to the meet the needs of the various types of landholders. One of these packages is presented in Table 15.6. It can be seen that a number of incentives are thought to be relevant to more than one of the groups.

Byron (this volume) questions the effectiveness of tree-planting subsidies, arguing that farm forestry is enabled if specific key conditions are met, including: secure rights to land; secure rights to tree crops; a viable production technology; capacity for plantation protection; and availability of timber markets. It can be argued on the basis of the studies reviewed in this chapter that the majority of landholders in sub-tropical and tropical eastern Australia do not perceive that the pre-conditions for farm forestry development as described by Byron have been satisfied. This suggests that Australian governments should place greater priority on removing structural impediments, relative to direct interventions to promote planting.

Table 15.6 Recommended programmes to support various types of landholders in developing farm forestry in North Queensland

Group number and name	Recommended farm forestry supports
1. High intensity	Provide information about multiple purpose plantings Provide tax breaks/incentives and rate reductions
2. Retired professionals hobby farmers	Continued Community Rainforest Reafforestation Program scheme (provide labour, information and organisation for planting activities), develop networks
3. Progressive second generation	Provide advice about plantings Provide tax breaks/incentives and rate reductions
4. Traditional	Develop options for short rotation plantations and annuity schemes
5. Experienced/ comfortable	Develop options for short rotation plantations and annuity schemes

16.6 CONCLUDING COMMENTS

Studies of landholder attitudes to farm forestry in tropical and sub-tropical eastern Australia have revealed that landholders support tree planting on private land for environmental and personal reasons. However, adoption of farm forestry for commercial timber production is to some extent limited to particular sections of the rural community with a strong personal interest in tree growing and low reliance on farm income. Most landholders appear unconvinced of the financial viability of commercial timber production on private land and are concerned about the stability of the regulatory environment for farm forestry. There is a danger that if farm forestry is linked in the minds of the 'real' farmers with the ex-city landholders this could reinforce the attitude that farm forestry is not real farming and should remain the province of State government agencies and private corporations.

Results of the studies described in this chapter reinforce the conclusions of Byron. While most landholders agree with the proposition that governments should do more to support farm forestry, they are most interested in the government establishing a supportive economic and legal framework in which forestry development decisions can be confidently made. It appears that the level of adoption of farm forestry in Australia will depend to a large degree upon landholders being able to feel secure in investing in such a long-term activity. In particular the regulatory environment and markets for timber need to be stable and transparent.

ACKNOWLEDGMENTS

Thanks to Alison Specht and Lyndon Brooks for their assistance in developing the methodology and statistics used to describe landholder attitudes. Thanks also to David Smorfitt and Jean-Claude Eono for their work in carrying out the surveys of landholders in Far North Queensland. This work was funded in part by the Rainforest CRC, based at James Cook University, Cairns.

REFERENCES

Broome, G. (1993), Land Degradation, Conservation and Reforestation in the Wet Tropics: A Survey of Rural Landholders in the Atherton, Eacham and Johnstone Shires, unpublished B.A. Honours thesis, Townsville Qld: James Cook University.

Byron, N. and Boutland, A. (1987), 'Rethinking private forestry in Australia', *Australian Forestry*, **50**(4), pp. 236–252.

Chamala, S. (1987), 'Strategies to overcome communication deficiencies in achieving sustainable farming systems', *The Journal of the Australian Institute for Agricultural Science*, **53**(3), pp. 164–169.

Commonwealth of Australia (1992a), *Compendium of ecologically sustainable development recommendations*, Canberra: AGPS.

Commonwealth of Australia (1992b), *National Forest Policy Statement*, Canberra: AGPS.

Commonwealth of Australia (1997), *Plantations for Australia: The 2020 Vision*, Canberra: Department of Primary Industries and Energy.

Dunchue, H.L. (1990), 'Eucalypt woodlots, attitudes and adoption. Case study: Gippsland Victoria', in *Biennial Conference of the Australian Forest Development Institute*, Bunbury, Western Australia: Australian Forest Development Institute.

Dunchue, H.L. and Sinclair, R.L. (1995), Development of Commercial Arrangements for Farm Forestry, Consultants report, Wodonga, Victoria.

Eono, J.C. and Harrison, S.R. (1996a), 'Report on non-timber benefits of rainforest reforestation', Annual Conference of the Cooperative Research Centre for Tropical Rainforest Ecology and Management, Kuranda, Queensland: Cooperative Research Centre for Tropical Rainforest Ecology and Management.

Eono, J-C. and Harrison, S.R. (1996b), 'Local Government and Landholder Perceptions of the Benefits of Planting Rainforest Tree Species in the Queensland Wet Tropics', in A. Keto, (ed.) *World Heritage Tropical Forests Conference: Handbook and Abstracts*, Cairns: Australian Rainforest Conservation Society: 8.

Emtage, N. (1995), Landholders' Perceptions of Planting and Managing Trees, B.App.Sci. Hons. thesis, Lismore: Southern Cross University.

Emtage, N.F. and Specht, A. (1998), *Landholders' Perceptions of Farm Forestry in the Northern Rivers Region of New South Wales: Overview Report*, report to the Northern Rivers Regional Plantation Committee, Lismore.

Emtage, N.F., Herbohn, J.L. Harrison, S.R. and Smorfitt, D.B. (in prep.), *Landholders attitudes to farm forestry in far north Queensland: report of a survey of landholders in the Eacham, Atherton and Johnstone shires*, Cairns: Rainforest Co-Operative Research Centre, James Cook University.

Harrison, S.R., Sharma, P.C. and Lamb, D. (1994), 'Progress in farm forestry in the Obi Obi Valley', paper presented to the Farm and Community Forestry Seminar, Dorroughby Field Studies Centre, NSW: Sub-Tropical Farm Forestry Association.

Harrison, S.R. and Sharma, P.C. (1995), 'Reforestation with high-value timber in Australia', in *GIS 95*, Vancouver, pp. 675–682.

Harrison, S.R., Eono, J.C., Herbohn, J.L. and Sharma, P. (1996), 'Attitudes to tree planting and assistance schemes by Queensland landholders', in *Managing and Growing Trees Training Conference*, Vol. 1 pp. 127–136.

Howden, P., Vanclay, F. Lemerie, D. and Kent, J. (1998), 'Working with the grain: Farming styles amongst broadacre croppers', *Rural Society*, **8**(2), pp. 109–126.

Kaine, G.W. and Lee, J.W. (1994), *Patterns in Innovation: An Analysis of the Adoption of Practices in Beef Cattle Breeding*, Armidale, New South Wales: The Rural Development Centre, University of New England.

Specht, A and Emtage, N.F. (1998), Landholders' Perceptions of Farm Forestry in the Northern Rivers Region of New South Wales, report to the Northern Rivers Regional Plantation Committee, Ballina NSW.

Stewart, A. and Reid, R. (1994), 'The design and development of agroforestry and farm forestry options to match farmer needs in the Otway ranges', in *Faces of Farm Forestry*, proceedings of the Australian Forest Growers conference, Launceston, Tasmania: Australian Forest Growers.

Vanclay, F.M. (1997), 'The social basis of environmental management in Australia', Ch. 2 in S. Lockie, and F.M. Vanclay (eds), *Critical Landcare*, Wagga Wagga: Centre for Rural Social Research, Charles Sturt University, pp. 9–27.

Vanclay, F., Mesiti, L. and Howden, P. (1998), 'Styles of farming and farming sub-cultures: appropriate concepts for Australian rural sociology?', *Rural Society*, **8**(2), pp. 85–108.

Wilson, S.M. and Yannick, D.T. (1995), *Trees on Farms: Survey of Trees on Australian Farms: 1993/4*, ABARE Research Report No. 95.7, Canberra.

16. Keys to Smallholder Forestry in Developing Countries in the Tropics

R.N. Byron

Most forest plantations are developed on large scale, by governments and corporations, in developed countries as in Australia. Yet this chapter argues that smallholder plantation forestry (including farm woodlots and agroforestry) is not an anomaly which requires special governmental assistance schemes and subsidies. Rather, tree-farming is a perfectly normal and common economic undertaking for small farmers and investors, wherever particular conditions are met and obstacles are not placed in the way. This chapter is based on experience in many developing countries of Asia and Africa, as well as Australia, North America and Europe. The analysis covers over 20 contexts in which farm forestry has taken off successfully, and over 40 where it either has failed, or has not been tried. By identifying the preconditions for success, R&D efforts or policy interventions can be focused to remove the key impediments and disincentives. Many previous 'assistance schemes' have been ineffective because they took a partial or inaccurate view of smallholder decision-making, which failed to recognise explicitly the interconnections between production, markets and policies.

16.1 INTRODUCTION

Farmers are commonly believed to be uninterested in trees, even hostile, and widely blamed for deforestation and land-clearing, especially in tropical developing countries. Yet farmers all over the world can and do plant and manage trees – whenever they perceive that to be more useful and beneficial than alternative activities with their land, capital and labour resources. Rather than presume that farmers will only become interested in forestry if offered some special inducement, there is some research and casual observation to suggest the opposite. One might well ask 'Why is it that *most* farmers and smallholders are *not* managing trees?'

Much can be learnt by studying the conditions under which farm forestry has arisen spontaneously, or after some market or policy intervention, compared with contexts in which it has failed, or simply never been attempted. But to begin, it is necessary to recognise that there has been a consistent bias against smallholder forestry in most countries (the most obvious and important exceptions being in Scandinavia) in regard to technical support, market structures and government policies.

There has been relatively little research or even field experience with smallholder timber plantations in tropical developing countries such as Indonesia and particularly about plantations on degraded and difficult sites such as imperata grasslands. Much of the silvicultural research has been in large-scale field trials, such as reported in Turvey (1994a,b). In considering obstacles to farm forestry, it might be important to assess how much of this information is potentially relevant to smallholders, and how readily it is accessible to them.

Techniques for large-scale establishment of (usually exotic) timber trees (such as pines, acacias and eucalypts) are now reasonably well known, although there is still substantial room for improvement in basic knowledge, and especially in the implementation of what is already known. Major areas for R&D are: species selection for a range of products, including for local uses, and genetic improvement; the use of pioneer species as 'nurse crops' for the establishment of longer-term, more valuable tree species; improved nutrition, especially at relatively low costs, such as through micro-symbionts; and improved fire protection.

But to what extent are these 'industrial' techniques involving mechanical site preparation, herbicides and so on applicable for, and adoptable by, *smallholders*? Are there any particular reasons why they would generally be unable to apply these techniques or simple modifications of them? It should be recognised that there are some fundamental differences between smallholders and plantation estates, most notably their capacity to invest, particularly in long-term investments, security of land tenure, and degree of influence or control over markets, prices, transport, etc. As a result, there may be fundamental differences in objectives, which inevitably lead to the choice of quite different processes and techniques. Estates may choose to maximise yield per hectare (in order to ensure supplies to their factories), while smallholders seek to maximise the cash return per unit of labour input. Smallholders will therefore not just mimic the techniques developed by plantation researchers:

- They are more likely to choose mixtures of multi-purpose species than even-aged mono-crops, for example, and can manage this great complexity more efficiently than a large company.

- They can manage small land areas much more intensively with household labour, than an estate could afford to, with a paid labour force.
- They can produce small quantities of specialty products (which may attract premium, boutique prices), rather than large quantities of bulk commodities.
- They can learn key insights from the expensive research undertaken by the estates (such as the imperative of weed control – no matter how it is best achieved – the importance of using the best germplasm, and the responses to different levels and types of fertiliser).
- They may well have great advantages over estates in protection from fires (due to population density, settlement patterns, site clearing and 'self-help' vigilance).

These factors may well compensate for any diseconomies of scale that smallholders may face. There are many reasons to think that most of the limiting obstacles to smallholder tree-planting are socio-economic and institutional, rather than silvicultural. Yet silvicultural research continues to receive more attention from foresters than the socio-economic factors. In this chapter, the focus is towards understanding the people first, before considering the role that trees may or may not have in their livelihoods.

16.2 WHO ARE THE SMALLHOLDERS – THE POTENTIAL TREE FARMERS?

Rural people in tropical developing countries have a variety of cultural attributes, skills and a diverse array of activities or 'household livelihood strategies' (that is, ways they allocate their limited resources, mainly labour) to generate sufficient to meet their needs and hopefully improve their 'quality of life'. These households generally have many choices of livelihoods – off-farm, on-farm or forest-related – and need to make decisions frequently (at daily, weekly, and longer intervals) about 'what to do?' In effect, they are repeatedly asking themselves, 'How can we best make a decent living for ourselves, given our skills, assets, resources, knowledge, inventiveness and constraints?'

Often the decision is easy – 'just do the same as last time we thought about this!' – but not always. The household may be under pressure from competitive markets for its produce, perhaps declining yields or increasing input costs, or perhaps a need to accumulate emergency cash reserves, as well as the desire to improve the standard of living obtained by the *status quo* strategy. On the positive side, they may have heard of new products, new technologies or new employment opportunities. In the forestry development context, various questions arise: How does each household choose to use/manage trees to gain particular goods and services to meet their household goals; (those goals are probably, but not solely, income-related; others may include prestige, social acceptance, stability, or conflict-minimisation)? Which products do they choose to pursue and why? How do their neighbours (and neighbours' choices) influence these choices? How are all of these influenced by government activities, (the implementation or otherwise of decisions and policies made at national or sectoral levels)?

Small farmers' use and management of the existing forest may include: destruction of primary or secondary forests for conversion to agriculture or pasture; establishing plantations or tree crops; enriching remnant forests with 'desirable' species (and just what is 'desirable' and why?); or just maintaining existing trees – with minimum effort, impact, and thus benefit. Often there are cleared lands, or former forests, perhaps now degraded and 'infested with weeds'[1] which potentially could be converted to plantations of timber trees, fruit trees or tree-crops such as rubber. There are countless examples across Asia (e.g. Ohlsson and Byron, 1988a,b; Byron and Ohlsson, 1989) where the mix and structure of multi-use trees selected, planted and managed by farmers has not been recognised by forestry professionals who were trained to think only in terms of monocultures for wood production. The economists' perspective is that the household decision-maker, seeking to optimise the household's living standards, responds to the relative prices of inputs and outputs which determine eventual 'profitability' or reward per unit effort. This is, of course, also subject to social factors such as cultural acceptability and status.

16.3 THE HOUSEHOLD'S DECISION ABOUT WHETHER AND WHAT TO PLANT

If land is relatively abundant and labour is the household's principal resource and source of income, a small-farmer's household livelihood strategy is to maximise the expected net benefits of the land-based activities, in terms of returns per unit of labour input[2]. An activity will only be undertaken if the

---

1 In economic terms, a weed is a plant for which there is no economic use, or which is in the wrong time or place, yet exactly the same plant can be a *resource*, if there is a use for it. The same applies to animal 'pests'; to a cultural group who do not catch and eat certain animals, they are a pest (e.g. Muslims' view of wild and feral pigs in Southeast Asia). Yet for those groups who do use them (e.g. non-Muslims), the same pigs are a food resource. Such value-loaded and context-specific terms as 'weed' or 'infestation' should be used cautiously.

expected net returns per labour input (adjusted for effort/danger) are higher than the alternatives (in other agriculture, forestry or non-farm activities). The choice of which activities to undertake, and how (which process) will be to maximize this objective, which can be defined as

Expected Net Benefits = Expected Quantity x Expected Price for each product, *minus* Production Costs (any hired labour, rented land, purchased inputs), *minus* Marketing Costs, per unit of household labour input or 'effort'.

The household or individual will retain existing trees, encourage or permit them to regenerate, or establish new trees (perhaps with different species in different figurations) according to the following rule of thumb: *Do the perceived total expected benefits of having these trees here outweigh the perceived costs?* In weighing up the pros and cons, we can list all the factors that are likely to tilt the balance in favour of trees, by either increasing the expected benefits or reducing the costs of doing so.[3]

Factors that Increase the Perceived Total Benefits

First, it is necessary to address the significance of 'perceptions'. If the prospective tree-farmer does not believe that he has security of tenure (meaning a very high probability that he will control, and benefit from, the disposition of the produce of his labour), then he will behave as if the expected values of future benefits are near-zero, no matter how great the physical output may be. Clearly an initial pre-requisite for retention *or* re-investment in trees, is a perception of reasonable security of tenure over the trees or land; without this, it does not matter how high the yield or the output prices or how low the production costs, the investment is unlikely to be worthwhile.

If the use of the forests is of doubtful legality, the forest user will factor into calculation of the expected gains, an allowance for the probability of being caught, and the size of the fine or bribe that would have to be paid. Thus the

---

[2] This is somewhat different to the objective of the non-operating owner who supplies capital and/or land, and hires labour, whose objective is presumed to be to maximize net returns per unit of capital invested or net returns per hectare. In the case of land shortage and labour abundance, maximizing returns per hectare would be rational, but this is not typical of most small-farmers, especially in developing countries.

[3] *Commercial Farm Forestry in Australia*, RIRDC (1998) lists about 80 factors which can be 'impediments to adoption of commercial farm forestry' under four categories: cultural; economic; public policy; and market impediments. However, many of these are inter-dependendent, and a listing has little explanatory power. Thus in the following section, an economic formulation is proposed which can explain how and why these factors operate to affect smallholders' decisions, either individually or as a complex interactive system.

intensity of policing and the legal context will influence the expected *net* gains from any possible course of action.

The lifespan of the investment (the cutting cycle of a natural forest or rotation length of the plantation) will affect the perception of security, as well as the risk that the physical outputs will be reduced through some physical damage (such as storms, fires, pests and diseases) or the value of the products reduced through lower selling prices or higher marketing costs.

The other important perception of a decision-maker is whether any given change in market conditions is considered to be just a short-term anomaly or hiccup (in which case it might pay to cut trees as rapidly as possible in response to higher market prices for logs, for example) or the start of a long-term trend (in which case the smallholder could rationally respond to exactly the same external stimulus, by retaining as many trees as possible now, buying up additional forests or harvesting rights from others, and planting as fast as possible for future markets). Analysts cannot predict *in advance* whether the exogenous change will give rise to a destructive, or conservative, response without knowing the perceptions and expectations of decision-makers. This is extremely important in the debate about the potential role and impacts of marketing non-wood forest products. *Ex ante*, it is difficult to predict whether increasing commercialisation and higher prices for non-wood forest products (NWFP) will lead people to:

- Over-exploit beyond sustainable levels, which often seems the case.
- Ration the limited supplies, releasing small quantities at high unit prices, rather than flood the market.
- Establish plantations and domesticated sources of supply.

In practice, all three responses often can be seen simultaneously. What are the positive signals or stimuli that will influence farmer behaviour? The most obvious are the *direct pecuniary benefits,* derived from either tree products or from non-timber products. Changes that will stimulate forestation by increasing the direct income from *tree products* include:

- Higher log selling prices (legally or otherwise), whether this is due to a change in government policy, creation of additional new markets, improved marketability (such as the ability to sell small quantities occasionally, or to sell lower grade products which were previously non-commercial, etc).
- Lower transport costs to market, or lower costs of delivering inputs, because of improved infrastructure (new roads or upgraded facilities).
- Higher physical yields.

Changes that stimulate forest management and plantations by increasing the direct income from *non-timber tree products* may include:

- Higher net prices due to increased consumer demand (more consumers or consumers with more money to spend, or increased prices of substitutes, or changed consumer tastes and preferences).
- Improved market access – elimination of middlemen, monopolies etc;
- Lower transport and marketing costs (improved techniques, better infrastructure).
- Enhanced yields of crops and livestock as a result of the trees' presence.
- Reduced attractiveness of alternative land-uses (e.g. the elimination of agricultural subsidies and price support schemes, or the effect of high labour costs on alternatives which are more labour-intensive).
- Reduced risks (climatic, market, pests and diseases, eviction).
- Reduced rate of time preference (as reflected in prevailing lower interest rates).
- Increased yields of the non-timber tree products themselves.

It is vital to recognise that increasing biological yields of the target product is but one way, often the least critical and useful, of increasing the farmer's net income from tree-farming. Yet this has consistently been the target of silvicultural research purporting to help farmers. While biological research to increase yields can occasionally generate financial gains, the probable gains are likely to be small unless the marketing conditions are also right.

In addition, there are numerous *direct non-pecuniary factors* which can stimulate forestation, such as the prospect of significantly reduced local (on-farm) erosion[4] or the maintenance of supplies of NWFPs for subsistence use.

Finally there may be significant *indirect benefits* which occasionally stimulate reforestation in the public interest, such as for watershed benefits and wildlife retention. From a personal viewpoint, participation in forestation, whether privately or as part of a community collective action, can bring benefits such as increased social acceptance, respect and status. These become increasingly more difficult to quantify, and perhaps increasingly doubtful in terms of their significance to most farmers' decision-making. But they do need to be included and may be decisive in specific instances.

Factors which Reduce the Perceived Costs of Retaining or Growing Trees

The most obvious are the direct savings in production costs through:

[4] This applies where farmers actually do perceive eroison as an issue: in many places they do not, even when outside experts see serious land degradation problems.

- New technologies and high yielding varieties.
- New management techniques (such as improved species-site matching).
- Capital subsidies (e.g. credit schemes) *if* capital costs are a significant constraint.
- Plant subsidies (free nurseries, etc.) *if* planting materials are a significant constraint.
- Improved knowledge, extension and training, *if* they are a significant constraint.

These are the areas in which most R&D and incentive schemes have operated, although often without first confirming that they are addressing the crucial, limiting factors. For example, it may sometimes be more beneficial to focus on reducing the opportunity costs of alternatives, e.g. by removal of subsidies or price supports for competing agricultural crops.

16.4 UNLOCKING THE POTENTIAL FOR SMALLHOLDER FORESTRY

Institutional economics suggests that individuals and households will combine to take collective action when the expected benefits outweigh the expected costs of doing so. Community action is certainly not costless – there are transactions costs in meetings to attend, lobbying, cajoling, enforcing – and direct or opportunity costs in that the communal decisions may preclude some action that would be directly, privately profitable. On the other hand, there may be courses of action which are only possible with the consensus of neighbours. For example, if most of one's neighbours light fires in the grasslands every year, it is most unlikely that any individual would risk a major effort in tree planting or maintenance. The examples from Madura and West Kalimantan, Indonesia (see Boxes 1 and 2) illustrate how small changes in local institutions, with minimal transactions costs, can result in major changes in land use, with gains in both farmers' incomes and environment.

Any policy intervention that reduces the transactions costs of convening group action, or increases the effectiveness of such group action (such as shared goals and culture, increased community cohesion, traditional leadership or authority figures) may well 'tip the balance' towards making collective smallholder forestry more viable and attractive. The policy questions then become: What different form of institutional arrangements, social organisation or economic relations could alter the choices of the majority of household decision-makers, to increase the net gains they perceive from managing (retaining, regenerating, re-planting or enriching) the forest. What then are the limiting constraints to the widespread application and adoption of smallholder, tree-based systems, particularly in developing countries or where such plantations can only be created on degraded or inherently low-productivity lands?

Box 1 Acacia auriculiformis plantations on Imperata grasslands in Madura

Kombangan Village, in Bangkalan District, Madura in East Java received a '*Kalpataru*' (environment award) from the government of Indonesia in 1993. They had developed a system utilising the formerly unproductive *Imperata cylindrica* grasslands to grow *Acacia auriculiformis,* in a part of Madura known for its dry and infertile soils.

Farmers began to cultivate 5ha of *A. auriculiformis* in 1972. The growth rate was so good that in 1980, 857 farmers participated in cultivating a total of 350ha. They cultivated the grassland intensively and planted *A. auriculiformis* at a spacing of 1 m by 1 m, applied manure to each seedling and intensively maintained the trees for two years. Animal grazing in this area was forbidden. Each farmer has been free to manage his woodlot and to harvest/sell *A. auriculiformis* wood without any government charges or interference. Their only responsibility has been to maintain their land.

At four years after planting, a thinning is done and final crop trees are ready to harvest any time after eight years when the tree diameter is 20 to 30cm. The price of *A. auriculiformis* six years old in the field in 1989 was Rp 10,000 ($US5) per tree. In 1994 the price of *A. auriculiformis* wood 10 –12 years old in the field was Rp25,000 ($US10) per tree. In general, 30% of the trees is used for lumber, 30% is used for raw material for furniture and 40% is used for firewood. The branches are usually left in the field by the buyers and could be sold by the farmer as firewood. The price of sawn timber from mature *A. auriculiformis* in the market ranges from Rp400,000 to 500,000 per cubic meter. Some of the wood is sold as far away as Jakarta.

Growing *Acacia auriculiformis* on degraded lands in East Java has generated a good income for the farmer and enhanced the environment. Marketing of the product has been secure and easy. This system could readily be extended to other grassland area and different perennial species.

Source: Dr. R. Sjamsul Arifin, Chief of Geger Sub District, Bangkalan District and Dr N. Gintings, Forestry Research and Development Agency, Bogor.

There is ample evidence that smallholders *could* establish and even protect small woodlots, but what we have not yet satisfactorily addressed is the conditions under which they would want to, as compared with choosing not to bother. Potential tree-farmers become active once they believe that the net benefits to themselves (not just to the State or their community) will exceed the best alternative use of their time, labour and effort, after taking into account the prospective markets, the likely yields, all the costs, and the risks (markets, biological, climatic, theft and expropriation, and fires).

Box 2 Village level regulations to protect trees from fires

Fire is one of the major obstacles to reforestation especially in *Imperata cylindrica* grasslands. For example 2,000ha of plantation in North Lampung and 2,000 ha of plantation in West Kalimantan, were burned in the 1994–5 El Niño period, and up to 40,000ha of plantations was reported destroyed in the 1997–8 El Niño fires. There are several reasons why people burn the grassland: the farmers believe that grassland harbours pests; people want to get the young grass-shoots for their livestock or for hunting the deer; as well as accidental fires (for example a smoker throws a cigarette stub into the dry grass) and occasionally, arson.

One village in Sanggau District, West Kalimantan, tried to formalise tribal custom to protect their trees from fire. For a number of years the people in Sanjan felt that there were real economic prospects to growing trees, but recognising that the grasslands burn every year, realised that tree-planting would be a waste of time, money and effort. Consistently, fires had damaged any farmer's plantations, even though there were supposed to be regulations to prevent fires.

Eventually, Sanjan people invited neighbouring villagers to a meeting to discuss the disadvantage of fire, and what they all might gain *if* the fires could be stopped or controlled. They eventually drafted their own set of rules.

If a farmer intends to burn grass or shrubs in his field, he must advise all adjoining farmers. On the day of burning, all the farmers in that area gather to prevent the fire from escaping. If the farmers can't manage the fire collectively, nobody is penalised. But if a farmer burns grass or shrubs without informing the neighbours, and the fire escapes and burns someone else's plantation, then that farmer is penalised as follows. If the damage is to a young plantation, the farmer has to re-establish a new identical plantation. If the plantation is already established (diameter more than 10cm), he should pay compensation of Rp50,000 ($US20) per tree.

This regulation was signed by the key person (*ketua adat*) from each sub village and also signed by the Chief of the Sub District. In the five years since the villagers established the new regulation, only one fire has occurred. That farmer confessed and paid all his dues based on the new regulation. Because all the farmers know the consequences of fire, and because the regulations were developed by the group members themselves, the group enforces and obeys the regulations very well.

Between 1980 and 1998, I conducted research analysing why there was an explosion of farm forestry in parts of Bangladesh, India, Viet Nam, and Southern China, yet farmers were rarely active in tree-planting on *Imperata cylindrica* grasslands of southern Borneo (Kalimantan). A parallel situation is that farm forestry is an integral part of the farming system in Scandinavia, New Zealand and the southeast of USA, but still is considered a rarity in many parts of Australia and Canada.

The conditions under which smallholders and farmers will establish tree plantations can be described as like 'a door with many locks'. The requirements are not necessarily sequential, and can be interdependent, but we need a framework for assessing which are the limiting factors – the bottlenecks – and where the greatest potential gains from intervention might be. Much of the research and literature has focussed on finding a 'key' for one or two of the locks (such as improving the biological productivity of the silvicultural system, or providing protection or insurance against fires), but if the metaphor is correct, it is futile to just open one or two locks on the door, if another three remain locked.

The 'locks' or impediments to farm forestry, when generally defined, are remarkably universal. They include secure property rights to land and tree crops; a viable production technology; capacity for crop protection; and adequate markets. From an analysis of all the situations studied, in over 30 countries as diverse as Sweden and Vietnam, looking at the common features of where farm forestry occurs and where it fails, this is the conclusion: *unless these four barriers are simultaneously overcome, widespread tree-planting is unlikely to occur. Conversely, successful sustainable tree-farming does exist whenever all four conditions are met simultaneously. No one of them is a sufficient condition for success, but all are necessary – failure on any one can be a guarantee of failure overall.* These requirements will now be detailed further in the following section.

Land and Crops Tenure

The 'need for security of tenure' has generated much excellent analysis, since the issue became widely recognised as a critical constraint in the early 1980s (Fortmann and Bruce, 1987). Although initially envisaged in the terms familiar to Western analysts (e.g. Freehold Title, Deeds) a rich variety of forms of land and tree tenure, vested in individuals, households or communities, has now been described. Certificates of Stewardship have been issued in the Philippines since 1978; Thailand's STK program granted moderately secure bundles of tenure rights to people whose farms may have been formerly (and formally) National Forest Reserve.

Although now it seems so self-evident that security of tenure is a pre-requisite for tree-farming, we have explored some apparent anomalies. First, there are isolated cases where the potential profitability of the tree crop is so great, that the grower considers it worth the risk of loss or eviction. Coco production in Bolivia is the closest example, but excluding such illicit high payoff short-payback crops, it is implausible that trees will be planted without perceived security of tenure over either the land and trees, or just the trees and their products. Ohlsson and Byron (1988b) document a case in Bangladesh where seven families had tenurial claims over a single mango tree, including the one who planted the tree, the current owners of the land, the previous owners, the family engaged in marketing the fruit and the village sawmiller who had a 'lien' on the trunk when it was ready for harvesting. Despite this complexity, tenurial arrangements were secure and well understood, the management of the tree was not problematic. Yet it is easy to see how the 'rights' could break down, and this would almost certainly lead to the demise of the tree.

Perceptions of security may bear no relationship to the legal status – the existence or otherwise of pieces of legal parchment. For example, in parts of northern Thailand certain indigenous (hilltribe) people have no legal rights to occupy, use and benefit from forests, yet behave as if they did, because they are confident that the King would never allow them to be evicted. Conversely, in Vietnam before Doi Moi reforms in 1989, peasants were given tenure documents, but behaved as if there was zero security of tenure because they feared the government would later renounce the certificates and re-expropriate the lands, once they had done all the hard work of making the 'wastelands' productive again. Perceived security of tenure is a necessary condition for tree-farming, but it is only the first of the locks that must be opened.

Viable Production Technology

Not surprisingly, biologists, agronomists and silviculturists consider their potential contributions through a new biological breakthrough – a new species or provenance, a more effective fertiliser regime, inoculation with microsymbiants, clonally propagated HYV germplasm – will hold the key, through increasing physical productivity per hectare, reducing costs or reducing disease risk. Many such important breakthroughs have been made, and more are needed, but it is important to understand the context. If all the other locks have not been opened, no amount of technological improvement will help, or be adopted. It is possible that technical research, disconnected from the socio-economic context, may even have unexpected or perverse effects (see Box 3).

Crop Protection Capability

Of course no farmer will invest in a long-term crop unless they are reasonably confident that the investment will survive to marketable maturity. If for convenience we subsume the risk of theft and expropriation under the heading of tenure (the security of ownership of a bundle of tradeable property rights), then this category can focus on the biological risks of pests and diseases, and climatic risks of fire, hurricane, hail, snow, drought, etc. The example of community-based fire control and protection from West Kalimantan, Indonesia (Box 2) illustrates how the ability to protect a plantation can be the limiting constraint, but once this is resolved, *provided* all the other keys are in place, the 'door opens' and smallholder forestry takes off. Initially in Australia, the unavailability of insurance against forest fires was a major impediment to farm forestry in some regions until the late 1970s when a system was developed by the Australian Forest Growers, and the premiums were affordable.

Box 3 Cupprascu in Acre, Brazil

This rainforest fruit is currently experiencing a boom in demand after its recent commercial 'discovery' and use in ice-cream. The physiology of the fruit is such that it must be eaten or processed within five days of maturity. Much of the existing supply is in forests that are some days walk from the nearest roads or markets. Much of the current production just rots where it falls, because it is not possible to get it to market.

Plant breeders have recently devoted considerable efforts to developing higher-yielding horticultural varieties of this tree, and concerted efforts have been made to persuade smallfarmers on the edge of the forests, to plant the improved varieties. To date most of these extension efforts have failed completely. It would be extremely naive to expect growers to plant improved varieties, that would produce even more fruit to rot; but if the improved varieties were in terms of reduced perishability, that could be very useful. The limiting factor for cultivation of this tree is not biological yield, but remoteness from markets, combined with its extreme perishability.

Perhaps the most effective way of increasing these fruit-producers' incomes is in improving their access to the markets, or the net returns they receive from sales. Eventually, the high-yielding varieties will be probably be grown in orchards near to markets and processing centres, and this will adversely affect the poor farmers in the remote areas, who collected wild fruit from the forests, and who were originally the stated clientele for the tree-improvement work.

Market Structure and Access

Farmers growing long-term cash-crops have understandable reservations about future markets. In all instances investigated thus far, wherever farm forestry is flourishing there are strong credible signals (sometimes reinforced by advance-sale contracts) that buyers will purchase the tree products. Conversely, we are aware of numerous instances where trees have been successfully grown, in the vague hope that buyers will emerge, only to find that either buyer-industries do not eventuate, or else the small-scale producers are excluded from participation in that market. In the case of Central New South Wales, smallholder plantation development ceased about a decade ago. Even though the owners have secure land tenure, a well-proven and viable production technology, the ability to protect plantations from fire (and insurance against fire, hail and wind) and there are viable processing industries located close by, there is one crucial limiting obstacle. Timber from small private plantations (including my own) cannot gain access to the pulpmill, particle board factory or sawmill because the industries have an exclusive contract with the State forestry agency which precludes them from purchasing any private timber. Clearly, what would be needed to re-invigorate farm forestry in this region is not improved technology or better fire insurance, but access to markets and the ability to sell 'harvesting rights' in advance.

Another illustration comes from Borneo, where aid agencies and a Finnish pulp/paper company have attempted to convince smallholders to plant *Acacia mangium* for a pulpmill planned for construction in about five years time. Although the farmers have reasonable security of tenure over the lands, and ability to grow and protect the trees, and they know how the trees grow there, they have no confidence at all that buyers really will appear when their trees are ready for harvest. They have no experience of growing pulpwood as a cash crop. Their alternative is to grow oil palm, where a market already exists, where they can see the buyers pass by every week, and they are confident that that can sell whatever they produce.

16.5 CONCLUSIONS

Some generalisations may be made, based on these and many other case studies observed around the world. *First, farm forestry does not work where all the 'keys' are not in place*! Famous orchards producing mangoes in Dinajpur, Bangladesh, which had been managed for centuries, were clearfelled in the 1980s due to transportation difficulties. Nothing had changed regarding the land tenure, the productivity of the trees, the prices of the fruit, the costs of labour, etc. The reasons the orchards become non-viable was that the road and river transport system had degenerated so much that it was no longer possible to transport such perishable fruit to the urban markets. At the same time, alternative new crops emerged, that were suitable for the drier sites above irrigation level, including tobacco, grains, and sugarcane. The trees were replaced by less perishable, and thus ultimately more profitable, crops. In many parts of south and south-east Asia, the technologies for reforestation with small plantations exist, but security of tenure remains unresolved and without community consensus, there will not be effective fire protection. Until farmers see prospective purchasers, and have assurances that they will indeed enter into transactions with thousands of micro-producers, and pay attractive prices for small parcels of pulpwood, and until farmers have reliable access to deliver their output to point of sale, extensive reforestation by smallholders remains unlikely.

Second, successful farm forestry does exist, where all the 'keys' are in place! Farm forestry with various exotic species has really taken off in many parts of northern, central and southern Vietnam. As noted by Midgley et al. (1996), the current profitability of tree farming systems in Central Vietnam results from a combination of factors. There is a national policy of allocation of use-rights (tenure) for degraded and barren lands, to households, for both homestead and gardens and for plantation forests. The production technology is relatively simple and well-known, as casuarinas, eucalypts and acacias had been grown in the area for 50 years, being easy to propagate, tend, maintain and harvest. Note that this alone is insufficient to explain the great expansion of their cultivation – without the policy reforms, the exotic trees could have remained just an amenity and protection species. The national government policy of economic liberalisation and transition to a market economy, gives households and individuals the ability to engage in private production and to sell their products; and has stimulated a great deal of economic activity, creating a market for the produce of rural households, such as casuarina and eucalypt poles and fuel, e.g. in brick kilns. New roads, sometimes built by local volunteers in recent years, provide access to the thriving commercial-construction centres.

The Center for International Forestry Research (CIFOR) and the International Development Research Centre of Canada (IDRC) have been documenting and assisting with smallholder plantations of Bamboo, Eucalypts and Chinese Fir on degraded lands in Southern China. Across five provinces, and an enormous diversity of social, economic and ecological conditions, it has consistently shown that wherever these conditions are met, farmers are already active in tree planting and management. The converse though is also true – where the limiting factor that people perceive as reducing the potential viability of tree-farming, can be isolated and redressed, smallholder forestry rapidly emerged.

Pulpwood plantations with *Albizzia falcataria* for the Paper Industries Corporation (PICOP) in the Philippines were highly successful, while the market existed, because smallholders had a secure and viable, labour-intensive production technology (Gregerson and Contreras, 1978). The PICOP model was widely cited as a success, but since the paper company went into receivership and the demand for pulpwood collapsed about 1990, the 'PICOP Miracle' of smallholder plantations has also collapsed.

Other well known examples such as Rubber Agroforestry in Sumatra, or the Homestead Gardens in Bangladesh and Sri Lanka meet the four conditions set out above, concerning security of tenure, a viable production technology, ability to protect the assets/crop, and confidence in the ability to market or utilise the products.

REFERENCES

Boutland, A., Byron, R.N. and Prinsley, R. (1992), *Directory of Assistance Schemes for Trees on Farms and Rural Vegetation,* Canberra: Bureau of Rural Resources and Greening Australia.

Byron, R.N. (1984), 'Peoples Forestry: a novel approach to forestry in Bangladesh' *ADAB News*, published by the Association of Development Agencies in Bangladesh, May-June, 1984, pp. 6–10. Reprinted in *Asian Agribusiness*, **1** (6), pp. 19–23.

Byron, R.N. and Boutland, A. (1987a), 'Rethinking Private Forestry in Australia: 1. Strategies to promote private timber production', *Australian Forestry,* **50**(4), pp. 236–244.

Boutland, A. and Byron, R.N. (1987b), 'Rethinking Private Forestry in Australia: 2. Strategies to promote trees on farms' *Australian Forestry*, **50**(4), pp. 245–252.

Byron, R.N. (1991), 'Cost Benefit Analysis and Community Forestry Projects', in D.A. Gilmour and R.J. Fisher, *Villagers, Forests and Foresters: the Philosophy, Process and Practice of Community Forestry in Nepal,* Kathmandu: Sahayogi Press.

Byron, R.N. and Ohlsson, B.O. (1989), *Learning from the Farmers about Their Trees*, Rome: FAO Forests Trees and People Program.

Fortmann L. and Bruce, J. (1987), *Forests,Trees and Tenure,* Madison: Land Tenure Center, University of Wisconsin.

Gregerson, H.M. and Contreras, A. H. (1978), *The Economic Analysis of Forestry Projects, Volume 2 – Case Studies,* Rome: FAO Forestry Dept.

Midgley, S.J., Byron, R.N., Chandler, F.C., Thinh, H.H., Son, Tran, V.H. and Hanh, H.H. (1996), *Do Trees Need Passports: A socio-economic study of the role of exotic tree and other plant species in Quang Tri Province, Vietnam,* Canberra: Australian Tree Seed Centre, CSIRO.

Ohlsson, B. O and Byron, R.N. (1988a), *Marketing Timber from Small Private Forests: Are there Lessons for Australia from the Swedish Experience?*, Working Paper No. 6, Canberra: Centre for Forestry in Rural Development, ANU.

Ohlsson, B. O. and Byron, R.N. (1988b), *Forestry in the Context of Rural Development in Viet Nam,* Centre for Forestry in Rural Development, Working Paper No. 8 Canberra: Australian National University.

RIRDC (Rural Industries Research and Development Corporation) (1996), *Commercial Farm Forestry in Australia*, Canberra: Joint Venture Agroforestry program, Rural Industries Research and Development Corporation.

Turvey, N.D. (1994a), *Afforestation and Rehabilitation of Imperata Grasslands in Southeast Asia: Identification of Priorities for Research, Education, Training and Extension*, ACIAR Technical Report No. 28, Canberra: ACIAR.

Turvey N.D. (1994b), *Plantation Development in Palau Laut, Indonesia* (Unpublished monograph prepared for ACIAR by FORTECH Pty Ltd, Canberra, ACT).

17. Lessons from Social and Community Forestry in the Tropics, with Particular Reference to India and the Philippines[1]

S.R. Harrison, A.S. Ghose and J.L. Herbohn

Social and community forestry in developing countries often has its origin in a desire to redress inappropriate historical forest policies which involved excessive clearing rates, undue concentration of property rights and poor sharing of resource rents. Important examples of community forestry are joint forest management, which has been widely adopted in India, and community-based forest management programmes in the Philippines. These programmes involve assistance to rural communities from government and non-government organisations, and have the potential to empower rural poor and provide livelihood benefits. While these programmes have undoubtedly had considerable success, they have also had notable limitations. There is a need for further development of community forestry programmes which target the rural poor.

17.1 INTRODUCTION

Rural communities in developing countries require timber for housing, fuelwood, farm timber and furniture. Loss of native forests makes this demand increasingly difficult to secure, and provides a strong imperative for reforestation. In general, low-income rural communities do not have access to timber from industrial and government forests, and hence rely on various small-scale forestry activities.

[1] Support from Australian Centre for International Agricultural Research for the project 'A Financial and Social Analysis of Adoption of Small-Scale Forestry Including Australian Timber Species in the Philippines' and from the Australian Research Council for the project 'Asset Poor Women in Development: Institutional Impediments and Lack of Social Dynamics: Evidence from Rural India' is gratefully acknowledged.

There is worldwide interest in co-management of native forests by government and community (Gilmore, 1999). Past initiatives in social forestry have been widely criticised in that while they may have produced large supplies of timber they did little to improve the livelihood of local communities (e.g. Soussan, 1988; Gadgil and Guha, 1992). Illegal logging and environmental degradation have often been associated with these programmes. Hence attention has turned to developing forest co-management systems which are acceptable to local communities, sometimes drawing on systems of property rights regimes which prevailed in earlier centuries. Alienation of common property forest resources and large-scale deforestation appear to have often been associated with colonisation and development of modern logging technology.

Social forestry had its origins in India, and India has also been prominent in development of forest co-management systems. In recent times, a timber shortage has emerged in the Philippines, where a variety of social and community forestry programmes have been introduced. Hence these two countries provide important case studies in social and community forestry. The next two sections of this chapter examine forestry management programmes for rural communities in these countries, including brief case studies from personal observation. Similarities and differences in the designs of these programmes are then examined, and some potential pitfalls in these programmes are identified, followed by concluding comments.

17.2 SOCIAL AND COMMUNITY FORESTRY IN INDIA

Social Forestry in India

Nath (1996, p. 95) noted that the National Commission of India in about 1975 coined the term 'social forestry', which he defined as 'forestry development activities on non-forest public as well as individually owned lands'. In this sense, social and farm forestry are overlapping terms. According to Shen and Contreras-Hermosilla (1995), although the social forestry programme was originally designed to increase tree planting on underproductive non-forest government and community lands, greater importance was accorded to farm forestry. State forest departments in India have come to realise that survival of protected forests depends on support of local people, rather than fences, fines and armed forces (Nath, 1996).

High population density and land subdivision limit land availability for forestry in India. MOEF (1988) divided small-scale forestry into eight classes – farm forestry, village woodlots, block plantations, and road, pond, rail and canal side plantations, and others – collectively called social and farm forestry. Species grown include acacias (especially *Acacia nilotica*), eucalypts, *Prosopis cineraria*, *Dalbergia sissoo*, poplars (under irrigation), casuarina, bamboo and kadamb (*Anthocephalus kadamba*).

As well as timber, forests provide farm tools, fuelwood and non-timber forest products (NTFPs) including fodder for livestock. Forestry is sometimes adopted because of the lower labour requirement than crops. This was apparently important after the 'green revolution' when more intensive cropping methods led to labour shortages. Small-scale forestry also provides soil protection and other environmental benefits, and crop protection in agroforestry situations, and is sometimes favoured to prevent occupation of farm land by squatters.

An interesting account of small-scale private forestry in India is provided by Dewees and Saxena (1997a,b). Australian eucalypts were rapidly adopted as a commercial crop in a number of states, particularly in the fertile north-west, in the late 1970s and early 1980s.

> Between 1981 and 1988, farmers in India were reported to have planted around 8.6 *billion* trees on private land. Many of the planting programs . . . were financed by the government, particularly with assistance from the World Bank, and from Canadian and Swedish donor agencies. The speed with which rural households adopted trees, primarily *Eucalyptus* hybrids, into their farming system, was truly phenomenal by any standard. The total newly planted area has been estimated to cover around 2.5 million ha. Most of these new trees were planted either in woodlots or around field bunds (Dewees and Saxena, 1997a, p. 210).

Trees were regarded as being more drought resistant and having less price variability than crops, and lower labour intensity of forestry freed labour for wage earning activities.

Social forestry in India met with mixed success. Most planting was undertaken by large farmers and absentee landowners, who could afford initial outlays and the wait for returns, typically of about six years for poles and longer for other products. Also, planting was dominated by upper-caste farmers, who 'are generally better educated and have more relations and friends in the bureaucracy. This gave them better access to seedlings, markets and extension advice' (Dewees and Saxena, 1997b, p. 260). Timber was directed to markets rather than to the needs of the poor, and pressure on native forests continued. There was limited planting on degraded land. Dewees and Saxena (1997a) noted that the supply of poles, fuelwood and pulpwood exceeded the market demand, leading to a price collapse. A contributing factor was subsidised timber from state forest departments. Aggrevating the price fall were the effects of trees on yields of adjoining crops, lack of marketing infrastructure and of government subsidies relative to crops, and perhaps exploitation by market intermediaries. Eucalypts fell out of favour, and in some cases plantations were uprooted. The situation for eucalypts is relatively favourable in West Bengal, where land is typically less suitable for cropping, and wood product markets have been stable.

However, it is considered that eucalypts 'while fast growing, generate few non-timber benefits, [and] do little to enrich soils, or rebalance hydrological cycles' (Hardiman, 1996, p. 109).

Joint Forest Management in India

Joint forest management (JFM) in India arose spontaneously out of efforts by forest dwelling communities to protect their access to resources. Prior to the arrival of the British in West Bengal, various tribal groups had communal access to forests. 'The north-eastern region of India is the abode of diverse hill tribes, who adopt jhumning (shifting cultivation) in which forests are cleared by slashing plants and shrubs, and the residue burnt to allow cropping with maize, vegetables and paddy' (Nath, 1996).

As noted by Tisdell and Roy (1996), natural resources including foresty land may exist as private property, state property (*res publica*), communal property (*res communes*) and open access property (*res nullius*). Western prejudices have favoured two of these types, viz. private property and state property:

> Before colonisation, the . . . varied caste access to different natural resources . . . ensured monopoly for a local group in the use of specific common-property resources, and was instrumental in the long-term sustainable use of these resources. The historical trend of a deteriorating resource base throughout India is traceable to the forced conversion of communal property into state and private property . . . (Berkes 1989, reported by Tisdell and Roy, 1996, p. 14).

Poffenberger and McGean (1996) have provided a lucid account of the historical erosion of common property forest resources and dispossession of forest dwellers in the 'Jungle Mahals' of West Bengal. Large, rugged inaccessible forest areas provided an ideal environment for Santhals, Bhumij and other tribal groups, living on forest products and shifting agriculture, whose belief systems were strongly grounded in the worship of nature. While the area was nominally under Mughal control, little attempt was made to extract revenues or exert political authority. In 1760, the district of Mindnapore near Calcutta was brought under British rule. The British periodically sent armed forces into the jungle in an attempt to exert their authority and extract land revenue. Tribal communities would withdraw to the jungle when attacked, and themselves raided new settlers on forest fringes.

A series of rebellions took place during the British Raj. Zamindars (landlords) were allocated large areas of land, and often replaced tribals with Hindu farmers, due to their greater farming expertise. Throughout the late 19th and early 20th century, many forest communities became increasingly indebted to moneylenders and tax collectors, causing mortgaging and loss of agricultural lands. Logging companies progressively encroached on the forests, and land was converted to agriculture.

Dispossession of tribal groups continued during the early post-independence period. The role of the state Forestry Department was mainly to protect the forests from illegal use by unauthorized persons. Foresters spent most of their time on armed patrol, hunting forest product poachers and evicting people encroaching on forest lands. The forests continued to degrade rapidly, and local people became increasingly aggressive towards the foresters, whom they generally considered corrupt.

In 1970, an Indian forester (A.K. Banerjee) found his trials with sal (*Shorea robusta*), eucalypts, teak and other species were being disrupted by villagers cutting fuelwood. He held meetings with villagers, offering employment and later a 25% share of sal timber plus rights to all minor forest products. The villagers ceased grazing and cutting in the trial area, and formed the first Forest Protection Committee (FPC) around the Arabari forests in 1972 (Poffenberger and McGean, 1996).

Little further development took place until the early 1980s, when various villages began forming volunteer patrols – often armed only with bows and arrows or spears – and fined pilferers or turned them over to forestry staff. In spite of conversion costs (for forest regeneration and protection), the number of FPCs increased rapidly, with new groups forming both autonomously and through government initiative. By 1988 more than 500 FPC had been formed covering an area of more than 70,000 ha. These groups worked closely with the forestry departments to regenerate and replant forest areas. A West Bengal Government Order in 1989 fully recognized the verbal agreement between the foresters and the local people, which among other things entitled FPC members to 25% of the final sal tree harvests.

According to Pande (1997, p. 120), 'Joint Forest Management (JFM) is an ongoing participatory process in which local communities and the State Forest Departments share products, responsibilities, and decision-making authority . . . So far 15 states (of a total of 25) . . . have committed themselves to try these new approaches . . . it is estimated that there may now be more than 10,000 forest protection committees . . . covering 1.5 million ha. of forest land'. Similarly, Sarin (1996) noted that about 10,000 villages in the states of West Bengal, Orissa and Bihal have begun protecting their sal forests. Sometimes development of these groups has been encouraged by state and local governments and NGOs, but in general they have been an autonomous movement. 'In the tribal belt of eastern India particularly, some of these groups have emerged organically and have functioned for generations, whereas others are only a few years old. The community organisations vary in size from a settlement cluster of less than a dozen households to numerous villages comprising thousands of families'

(Sarin, 1996, p. 165). Community forestry is confined largely to the upland areas, the better agricultural land typically being under private ownership.

Approximately 50 to 60 million Indians rely on forests for their subsistence (Pande, 1997). They derive lumber and fuelwood from the forests. As well, non-timber forest products (NTFPs) 'are derived from over 3000 plant species. These products include medicinal plants, edible plants, resins, gums, oils and fats, essential oils, tannins, dyes, bamboos, canes, fibres, grasses ...' (FAO, 1999). The World Bank has supported site-specific participatory micro-planning initiatives, and forest extension and education programmes 'are pursued aggressively by about 500 NGOs' (Pande, 1997, p. 115).

The West Bengal Joint JFM is considered to be a model of participatory forest management. Ren and Tisdell (1996, p. 402) comment that 'in villages where community forests have been established, biodiversity has been increased and the extent of poverty appears to have been reduced.'

Some Observations of Indian Tribal Villages

A visit was made by one of the authors to four tribal villages around Chandrakona in Medinipur district of West Bengal in September 1999. These villages had access to jointly managed forests, with rights to 100% of NTFP and to about 30% of the timber. Their reliance on small timber, forest grazing, and NTFPs such as sal and bidi leaves, mushrooms and native potatoes was apparent. Production of plates from sal leaves was a prominent enterprise, and the major source of cash income of one of the villages visited. Women were the main forest workers, with a female chairperson in one village.

17.3 SOCIAL AND COMMUNITY FORESTRY IN THE PHILIPPINES

From about 1960 to 1990, rapid deforestation took place in the Philippines, taking that country from one of the world's largest log producers and exporters to a net timber importer. Timber license agreements (TLAs) were 'given to well-connected businesses, politicians and powerful families [as] a tool for political patronage' (Vitug, 1997, p. 335). The typical license area was 40,000 to 60,000ha. Clear cutting was common, and there was little monitoring. The Aquino government on coming to power in 1986 ceased the issue of TLAs and adopted a loan-funded reforestation programme and community-based forestry.

Social Forestry in the Philippines

Various programmes of social forestry have been introduced in the Philippines. Notable is the national Integrated Social Forestry Program (ISFP), designed to 'maximize land productivity, enhance ecological stability and improve the socioeconomic condition of forest occupants or communities' (Sy, 1998, p. 9). Under this programme, 'a qualified upland farmer or forest community association or cooperative may be awarded a Certificate of Stewardship (CS). As part of the contract, the government allows the holder peaceful occupation, possession, and sustainable management of the designated area' (DENR, undated). The agreement is valid for a period of 25 years, and renewable for another 25 years. The farmer is given technical assistance, credit assistance from the Land Bank of the Philippines and assistance from NGOs. The occupant is required to develop at least 20% of the land for tree farming, protect and conserve forest growth, prevent and suppress fires, protect and preserve trees and other vegetation within 20 metre strips along rivers and streams, and refrain from cutting or harvesting naturally growing trees in adjacent areas (DENR, undated). In a sense, this programme legitimises some of the squatter activity, common in some upland areas. This programme is essentially one of promoting private or farm forestry rather than common property forestry.

Community Based Forest Management

In 1990, the Department of Environment and Natural Resources formally introduced community-based forestry as a flagship programme, which was tested for five years in upland and coastal communities (Vitug, 1997). The community-based forest management (CBFM) programme was then established under Executive Order No. 263 in 1995, as a national strategy to ensure the sustainable development of the Philippines' forest resources. The strategy 'is the organised efforts of the government to work with communities in and near public forests aimed to protect, rehabilitate, manage, conserve and utilize the resources. The CBFM program integrates and unifies all current people-oriented forestry programs of the government . . . ' (Sy, 1998, p. 9). The integrated social forestry programme now falls within the CBFM umbrella[2]. A Community-Based Forest Management Agreement (CBFMA) entitles participants to occupy, possess, use and develop forest land and its resources, for a duration of 25 years, renewable for another 25 years (DENR, 1998b). Local communities are organised into

---

[2] Other programmes coordinated within CBFM include the Community Forestry Program (CFP), Coastal Environment Program (CEP), Forest Land Management Program (FLMP), National Reforestation Project (NFP), Forest Sector Project (FSP), Low Income Upland Communities Project (LIUCP) and Recognition of Ancestoral Domains (Sy, 1998; DENR, 1998).

People's Organisations (POs) to participate in the programme. The Community Environment and Natural Resource Officer (CENRO) validates the application for endorsement with the Regional Executive Director of DENR.

Approximately 16 million Filipinos live in the uplands, placing heavy pressure on forest resources. CBFM recognizes that communities are in the best position to protect and manage the forests. Areas typically are in the range 500 to 1,000ha, and are designed to provide sufficient opportunities for livelihood before trees are ready for harvest. The areas are often expired or cancelled timber license areas, with the community comprising mostly former TLA employees. Labour inputs and rights to forest products are shared by many members of a community. Programmes are sometimes organised with the assistance of NGOs. Since residents of upland communities typically do not have sufficient capital to commence a forest enterprise, the DENR advances some of the operating costs (Vitug, 1997). The government may maintain a minority share of timber rights.

Some Observations of a Philippines Community Forestry Programme

Two of the authors visited the community forestry programme in the village of Alcoy in Cebu, the Philippines, in February 1999. The people's organizsation managing the programme is the KMYLB Agro-Forestry Cooperative. Project components include 95ha of reforestation, 30ha of agroforestry, 162ha of rattan and 70ha of assisted natural regeneration; the estimated development cost is 3.16 M pesos ($A1≅25 pesos).

The project commenced in 1986, with planting of mahogany, falcata and gmelina. Subsequently, various endemic species and some eucalypts (including *E. camaldulensis* and the native *E. deglupta*) and casuarina have also been planted. Two nurseries produce seedlings, with sub-nurseries near planting sites. Prunings are given to members for their own use, while thinnings have been harvested for fuelwood. Some earlier plantings are now ready for harvest; 30% of revenue will go to DENR, and 70% to the community.

There are 115 members in the programme, from one barangay, comprising husbands, wives and sometimes both partners. Each member must contribute one day of their labour per week, although some contribute two days. There are 400 households residing in the cultivable and disposable area. Many of the members have other land, typically one to five ha, growing corn and vegetables, and raising cattle, pigs and chickens, and selling produce in Cebu City. Some plant trees on their own land, usually along boundary lines.

The plan is to sell finished timber products, and there is an expectation that tools including bandsaws will be donated by an NGO. Barangay members possess some furniture making skills, and plan to obtain training in cabinet making. The co-operative owns a store which sells food and other basics, the profits from which are disbursed to members on the basis of spending. The net assets of the co-operative after allowing for debt are 700,000 pesos, excluding trees.

Alcoy is a model community forestry project, with high commitment and organizational ability, and financial recording of a professional standard. Visitors have come from Vietnam, Sri Lanka and various places in the Philippines to inspect the programme. The community holds field days, and members visit other barangays to make presentations on their programme. This community programme has received considerable external funding from international agencies and NGOs, and the ease with which it could be copied in other areas is difficult to determine.

17.4 SIMILARITIES AND DIFFERENCES IN SOCIAL AND COMMUNITY FORESTRY BETWEEN INDIA AND THE PHILIPPINES

There are a number of similarities in circumstances surrounding community forestry programmes in India and the Philippines. In both cases, the motivation is to improve the livelihood of low-income rural people who have little individual land resources, and to redress the environmental damage of excessive logging of native forests. A further objective is to encourage people to settle at a fixed location, and practice sound forest management rather than illegal logging or shifting agriculture. In both models, participants in community forestry may own and work some land of their own, such as for vegetable plots and woodlots. Forest grazing appears to be important under both programmes.

Table 17.1 summarises some of the key differences between Joint Forest Management in India and Community Based Forest Management Agreements in the Philippines. The availability of land for community forestry is greater in the Philippines, where all sloping land (over 20 degrees) is under the control of the DENR. As well, the generally higher land quality and rainfall in the Philippines allow faster growing and more valuable species to be used. Landrights of indigenous inhabitants (Ancestoral Domain) tends to be more of an issue in the Philippines.

NGOs have promoted CBFMA, whereas JFM has originated spontaneously from the people, with some support from state forestry departments. JFM is a fixed-share joint venture between government and the community, whereas community forestry under the CBFMA model is a joint venture between members and perhaps external investors. Outside finance is now being courted for CBFM, where 'communities can enter into partnership contracts with investors. With the entry of the private sector comes capital, technology and managerial expertise . . . some of these communities have existing micro-enterprises and livelihood activities (bamboo plantations, rattan supply) . . . All the communities need are partners who know the business and, of course, who have the capital to finance larger enterprises . . . Private individuals, associations, cooperatives and corporations may invest in CBFM sites.' (DENR, c1999).

In CBFMAs, government permission is required to harvest timber, but property rights to that timber lie mostly with the community. The stronger timber rights under CBFMAs provide greater incentive for value-adding in timber products. Under JFM, the government provides silvicultural skills and marketing power with respect to timber sales. In JFM, mainly women work in the forests, whereas under CBFMA the work is shared more equally between gender. People in JFM frequently take on paid agricultural tasks such as rice planting and harvesting on nearby or even distant farms.

Table 17.1 Comparative features of JFM in India and CBFMA in the Philippines

JFM in India	CBFMA in the Philippines
Initiated by local communities	Designed by government and NGOs
Government continues involvement in forest management, with only informal management by community	Forests are established and managed by the community, and this may be along formal business lines
Majority share in timber rights retained by government forest service	Government holds only a minority timber right
Informal arrangements for community inputs, particularly by females	Community members agree to a minimum weekly input
Community members have no prescribed individual shares	Community members have individual equity shares
Little input of external finance	Finance obtained from government, international agencies, joint venture partners and NGOs
Forest protection committee must include female members	Management committee membership not prescribed
Emphasis on assisted natural regeneration and utilization of native forest	Emphasis on both reforestation and assisted natural regeneration plus non-timber crops (raffia, agroforestry)
High community reliance on non-timber forest products	Low community reliance on non-timber forest products
Few community members own land	Many community members may own land
Aimed at supporting the livelihood of tribals	For non-tribals
Community members undertake considerable outside paid work	Little paid work outside the community is undertaken
The adoption rate has been rapid	The uptake has been slow

17.5 SOME PITFALLS IN COMMUNITY FORESTRY PROGRAMMES

The deficiencies in social forestry are well documented. However, it would be naive to think that community forestry programmes will overcome all of these problems. There is concern that JFM arrangements may fail to solve social problems or even lead to new problems (Buchy, 1999). For example, while it is required that there be two women members on the forest management committee, these females may have token membership with little influence on decisions. Even if women generate additional income from forest products, they may not have control within the family on how this money is spent, hence there may be little improvement in gender equity. Another concern is that property rights may not be equitably distributed, with some individuals or groups gaining superior forest use rights. For example, one village may gain access to a forest at the expense of another, hence establishing new elites alongside even more disadvantaged people. Forest productivity under JFM would appear to be rather low, due to use of slow growing species (particularly sal), and relatively limited soil fertility and rainfall in many of the areas. Silvicultural skills of villagers probably are limited. Officers of the Indian Forest Service are critical of JFM, arguing that it is impractical except in the special cases of West Bengal and Bihar where 'a number of schemes had met with success because of the charisma of individual forest officers or individual village leaders . . . JFM was seen to work in smaller villages with more homogeneous populations but not in the larger or more heterogeneous villages where there is greater scope for a clash of interests' (Hannam, 1999, p. 227).

The uptake of CBFMAs in the Philippines has not been rapid. It would appear that considerable transactions costs arise in the establishment of such programmes, and facilitation and funding support is necessary to launch new agreements. Vitug (1997) discussed a case study in Mindanao where rival politicians fought bitterly for control of the prized concession through two opposing cooperatives, delaying the CBFMA for two years, and noted the importance of insulating these programmes from politics. It may be that success of any agreement depends on people coming forward with strong leadership skills, and even then external support for capacity building is required.

This discussion suggests that there must be a continued search for the most desirable community forestry arrangements, and that programmes should evolve over time in response to perceived problems and local needs. Greater empowerment may be possible through value-adding with respect to timber (envisaged in the Alcoy community forest in Cebu) and in more commercialisation and marketing of non-timber forest products (examined by Mallik and Panigrahi, 1998).

17.6 CONCLUDING COMMENTS

Top-down forestry, planned and implemented by government, is not likely to be well accepted by rural communities in developing (or developed) countries. However, community forestry in which local communities have a genuine say in forest planning and an active role in forest management has great potential to improve the livelihood of rural communities, and to foster sustainable forestry systems in close sympathy with the needs and aspirations of rural communities.

India and the Philippines provide informative examples of community forestry. CBFMA arrangements in the Philippines are a conceptually ideal design by government; JFM in India is a community-driven design. There are considerable differences between the two approaches. The uptake of JFM has been spectacular, and while pioneered by the people it has since been embraced by government. The uptake of CBFMAs in the Philippines has been slower, and external financial and training assistance may be necessary for successful establishment.

Motivations are similar – rural poverty, deforestation and other environmental problems (such as erosion, siltation and flooding). Governments have come to recognize the importance of non-timber forest products in the livelihood of low-income rural communities, particularly minority groups such as tribals. In both CBFMAs and JFM, local communities have almost exclusive property rights to non-timber forest benefits, and this can form a major part of their income. The two programmes have somewhat different contractual arrangements with government. Under CBFMAs, forest management and forest benefits are almost exclusively the province of the community, c.f. in JFM the government retains a strong presence in forest management, and receives most of the timber. There remain a number of dangers with community forestry, in that new elites may develop while others lose traditional forest access rights.

REFERENCES

Buchy, M. (1999), 'Co-Management Models in India', Address to the *Forum on Co-Management of Forest Lands*, Ipswich: Australian Tropical Forests Centre, The University of Queensland.

DENR (Department of Environment and Natural Resources) (1998a), *Community-Based Forest Management: Communicator's Training Manual*, Manila: Community-Based Forest Management Office.

DENR (Department of Environment and Natural Resources) (1998b), *Growth and Yield and Economic Analysis of Selected Forest Tree Species*, Davao City: Ecosystems Research and Development Service.

DENR (Department of Environment and Natural Resources) (undated), *Processing and Issuance: Certificate of Stewardship,* Quezon City: DENR.

DENR (Department of Environment and Natural Resources) (c1999), Investing in CBFM Communities, Information Pamphlet, Community Based Forest Management Office, Quezon City.

Dewees, P.A. and Saxena, N.C. (1997a), 'Wood product markets as incentives for farmer tree growing', in J.E.M. Arnold and P.A. Dewees (eds), *Farms, Trees and Farmers: Responses to Agricultural Intensification*, London: Earthscan.

Dewees, P.A. and Saxena, N.C. (1997b), 'Tree planting and household land and labour allocation: case studies from Kenya and India', in J.E.M. Arnold and P.A. Dewees (eds), *Farms, Trees and Farmers: Responses to Agricultural Intensification*, London: Earthscan.

FAO (Food and Agricultural Organisation, United Nations) (1999), Forestry – Profile – India, http://www.faao.org/infosylva/VR/is_country.

Gilmore, D. (1999), Collaborative Management of Forests for Conservation and Development: Approaches from Around the World, Address to the Forum on Co-Management of Forest Lands, Ipswich: Australian Tropical Forests Centre, The University of Queensland.

Gadgil, M. and Guha, R. (1992), *This Fissured Land: An Ecological History of India,* Delhi: Oxford University Press.

Hannam, K. (1999), 'Environmental management in India: recent challenges to the Indian Forest Service', *Journal of Environmental Planning and Management,* **42**(2), pp. 221–233.

Hardiman, D. (1996), 'Farming in the forest: the Dangs 1830–1992', Ch. 4 in M. Poffenberger and B. McGean (eds), *Village Voices, Forest Choices: Joint Forest Management in India*, Delhi: Oxford University Press.

Mallik, R.M. and Panigrahi, N. (1998), *Non-Timber Forest Produce Collection: Benefits and Management in Orissa*, Bhubaneswar, Orissa: Nabakrushna Choudhury Centre for Development Studies.

MOEF (Ministry of Environment and Forests) (1998), *Annual Report 1997–98*, http://www.nic.in/envfor/report/9798/con_nat.htmp.

Nath, K.P. (1996), 'India's dwindling forests: a threat to environment and eco-system', in R.K. Sen and K.C. Roy (eds), *Sustainable Economics Development and Environment: India and Other Low Income Economies (LIEs)*, New Delhi: Atlantic Publishers and Distributors, pp. 90–98.

Pande, S.K. (1997), 'Sustainable forest management: an Indian perspective', in A. Johnson and P.B. Durst, (eds), *Implementing Sustainable Forest Management in India and the Pacific*, Proceedings of a Regional Expert Consultation held in Bangkok, December 12–15, 1995, pp. 109–121.

Poffenberger, M. and McGean, B. (1996), *Village Voices, Forest Choices: Joint Forest Management in India*, Delhi: Oxford University Press.

Ren, Z. and Tisdell, C. (1996), 'Developing community-based forestry in the uplands of Yunnan: dictates of environment and socio-economics', in C.A. Tisdell and J.C.H. Chai (eds), *China's Economic Growth and Transition: Macroeconomic, Regional, Environmental and Other Dimensions*, Economics Conference Monograph No 2, Brisbane: The University of Queensland, pp. 401–414.

Sarin, M. (1996), 'From conflict to collaboration: institutional issues in community management', in M. Poffenberger and B. McGean (eds), *Village Voices, Forest Choices: Joint Forest Management in India*, Delhi: Oxford University Press.

Shen, S. and A. Contreras-Hermosilla (eds) (1995), *Environmental and Economic Issues in Forestry*, World Bank Technical Paper No. 281, Washington DC: The Word Bank.

Soussan, J. (1988), *Primary Resources and Energy in the Third World*, Routledge Introductions to Development, London: Routledge.

Sy, M.U. (1998), 'Rehabilitation of natural logged-over forests: The Philippines scenario', *Canopy International*, November–December, pp. 3, 9–11.

Tisdell, C. and Roy, K. (1996), *Governance, Property Rights and Sustainable Resource Use: Analysis with Indian Ocean Rim Examples*, Working Paper No. 1, Economics, Ecology and the Environment, Brisbane: Department of Economics, The University of Queensland.

Vitug, M.D. (1997), 'The politics of community forestry in the Philippines', *Journal of Environment and Development*, **6**(3), pp. 334–340.

18. Institutional Impediments to Farm Forestry

S.M. Vize and C. Creighton

This chapter provides a brief review of farm forestry across Australia with emphasis on North Queensland, from the perspective of public agencies concerned with land management. A list of elements of successful farm forestry initiatives is identified. The chapter discusses a range of problems and suggests solutions to institutional impediments to progress.

18.1 INTRODUCTION

Farm forestry is viewed here as growing trees for profit on farms. This covers two types of activity:

- growing trees in plantations of various scales and configurations – that is in wood lots, alleys, timber belts and multipurpose shelter belts, wide spaced plantings integrated with pasture and even large plantations; and
- managing and harvesting naturally growing trees on private farmland (Cummine, 1995).

Timber production has always included a component obtained from private lands; however, timber production and marketing systems have up until now been focused on and driven by the operations of government managed State Forests and Crown Reserves. Also, silvicultural practices have largely been confined to these government operations, with farm forestry generally confined to harvesting with minimal management on most private holdings.

Farm forestry has emerged in Australia as an additional on-farm activity for a range of reasons including:

- mitigation of degradation, e.g. in areas such as groundwater management at salinity hazard sites;

- recognition of the potential income stream, as a diversified source of income additional to normal farm income;
- environmental stewardship, since a shift in ownership of Australia's farms has meant that many landowners seek a long-term, lower maintenance level crop that also meets their environmental values;
- privatisation and rationalisation of government control on forest industries, with Australia-wide initiatives such as the National Competition Policy. Many foresters foresee that competition will ensure timber will not be held at artificially low prices that have dominated Australian product to date; and
- enhanced farm management, including riparian reforestation, shelter-belts and windbreaks.

This chapter reviews institutional arrangements and impediments to farm forestry using North Queensland as a case study. Suggestions for policy development, activities, incentives and regulation accompany discussion of each of the impediments.

18.2 CURRENT INSTITUTIONAL ARRANGEMENTS AND SHORTCOMINGS

The three levels of government in Australia have specific interests in and responsibilities for forest management, with State and Territory governments having primary responsibility (NFPS, 1992, p. 1). At the national level, the *National Forest Policy Statement* (NFPS) has been developed to: 'ensure that private native forests are maintained and managed in an ecologically sustainable manner' and 'to expand Australia's commercial plantations of softwoods and hardwoods so as to provide an additional, economically viable, reliable and high-quality wood resource for industry...[and] rehabilitate cleared agricultural land' (Commonwealth of Australia, 1992, p. 5).

Other national strategies with relevance to farm forestry include the *Wood and Paper Industry Strategy* (WAPIS), *Plantations for Australia – the 2020 Vision* and creation of the *Natural Heritage Trust* (NHT). The Wood and Paper Industry Strategy recommended the formation of the *Farm Forestry Round Table* and the appointment of a National Farm Forestry Facilitator as an avenue for addressing impediments to farm forestry (Forests Taskforce, 1995). The Round Table was first convened in February 1998.

Plantation 2020 Vision Implementation Committee (1997) has developed a vision to treble Australia's plantation forestry estate by the year 2020. An estate of this size – integrating State plantations, industrial plantations and farm forestry plantings – will be large enough to support a fully competitive commercial industry to supply the plantation timber needs of Australia. There are strong links between the *Plantations 2020* and *WAPIS*, particularly in the area of addressing impediments to forestry development.

The *Natural Heritage Trust* is an integrated package of government assistance for environmental and ecologically sustainable natural resource management initiatives. The Trust replaces and builds upon a number of separate funding initiatives, such as Farm Forestry Program and Bushcare. The Farm Forestry Program commenced in 1992 following the launch of the National Forestry Policy Statement (AACM, 1996a, p. 90), and provides the main form of Commonwealth support to farm forestry. The focus of the Farm Forestry Program, which is managed by the Department of Primary Industries and Energy, is on forestry extension to landholders to encourage the establishment of farm forests and assist in their management (AACM, 1996a, p. 91).

Bushcare is administered by Environment Australia and builds on a number of initiatives, including One Billion Trees which established native species farm forests as a component of the overall revegetation target. One Billion Trees funding was used to develop extension and promotion of farm forestry as a means of integrating commercial and conservation benefits (Robins *et al.*, 1996). Bushcare incorporates this development, and a significant proportion of funds has been allocated to continue the promotion of farm forestry, with benefits for the management of biodiversity and other environmental values.

The States have a constitutional responsibility for landuse decisions, the development of the management policies for forests on public and private lands, and the delivery of most Commonwealth-funded initiatives. In Queensland, forestry issues are predominantly managed within the Department of Primary Industries (DPI), and the Department of Natural Resources (DNR). DPI-Forestry has the lead role in the support of farm forestry, and is involved in delivery of the Farm Forestry Program of the Natural Heritage Trust.

Institutional arrangements for forestry in Queensland have changed from a Forest Service, which at one time also included management of National Parks and related reserves to:

- a specific section within the Department of Primary Industries dealing with all forestry issues.
- the separation of commercially related activities from natural resources management activities into separate sections, still within the Department of Primary Industries.

- the current structure that has commercial forestry interests in the Department of Primary Industries – Forestry and natural resource management and recreation related activities within State Forests and other Crown lands in the Department of Natural Resources.

Post World War II farm forestry was promoted in North Queensland based largely on Kauri (*Agathis robusta*), Hoop Pine (*Araucaria cunninghamii*) and several non-native pines (predominantly *Pinus caribbaea*). Several of these plots still dot the landscape. There has been no silvicultural management in virtually all of these plantations, and lack of management, particularly thinning, has resulted in the plantations being 'locked up' with growth severely retarded.

The statewide *Tree Assistance Scheme,* managed by the DPI and then DNR operated for a number of years providing extension advice and some free trees to landholders. Debate has ensued as to whether the Scheme should continue in its current form, particularly following the development of private nurseries and the apparently unfair competition that the Scheme represents in terms of provision of stock.

The DPI and then DNR have also managed the *Community Rainforest Reforestation Program* (CRRP). This programme commenced in 1993 as part of a structural adjustment package in north Queensland, following cessation of rainforest logging in State Forests from Hinchinbrook to Cooktown. A similar scheme was initiated in the Mackay area following the cessation of rainforest logging there in about 1995. The CRRP is one of the few Australian programmes that targets high value species in relatively small lots on farms (2 to 20ha with preference in latter years for plots greater than 5 ha) Some statistics on the CRRP are presented in Table 18.1

The 1996 change in State Government and subsequent change in priorities within DNR led to the immediate cessation of the Mackay component. The far north Queensland programme was wound down to provide extension services, but no longer any subsidized establishment of farm forestry lots. The impetus for full commercialization of State Forestry operations was also reduced. The DPI-Forestry departmental climate of corporatisation and strong support for plantation forestry led to a statewide Joint Venture Scheme commencing in 1997. This Scheme aims at more 'industrial' farm forestry concentrating on monocultures of fast growing species in relatively large plots (not less than 10ha).

DPI-Forestry has also been successful in gaining Natural Heritage Trust funding to establish farm forestry marketing co-operatives across the state. This timely initiative has the goal of enhancing farm forestry through structural and institutional support and is welcomed by the farm forestry community. Growers from the Community Rainforest Reforestation

Program have already founded a marketing cooperative in North Queensland. The Natural Heritage Trust support will be used to further develop the co-operative and integrate the interests of a variety of forest growers.

Table 18.1 Community Rainforest Reforestation Program outcomes

Area under trees	2,000ha
No. of trees planted	About 2 million seedlings
Dominant species	*Flindersia brayleyana, F. schottiana, F bourjotiana, Acacia melanoxylon, A. mangium, Toona ciliata, Eucalyptus grandis, E. cloeziana, E. pellita, E. dunnii, E. pilularis, Elaeocarpus angustifolius, Grevillia robusta* and exotics such as *Cedrala odorata, Khaya nyasica* and *Khaya senegalensis* with in total about 200 species trialled in various plots
Est. survival rate	85%
Planting regimes	Regimes varied and included agroforestry such as wide spaced hoop pine with cattle grazing, rainforest plantings, mixed cabinet species plantings, eucalypt/ acacia plantings and single species plantings.
No. of participants	560
Main reasons for plot failure	At most, 15% of plantations are considered failures. The main cause for failure was cattle damage, with early failure in some plantings due to lack of landholder commitment to maintenance.
Farmer contribution	Farmers paid a planting levy of $600/ha and also provided all site preparation and maintenance costs once trees were established.
Landholder commitment	More than 70% still committed to farm forestry

Source: QDNR (1998).

Local Governments have responsibilities for local land use planning, development control plans and rating systems, all of which have implications for farm forestry. There are several tree planting support initiatives operated by Shire Councils in Queensland. The Wet Tropics Tree Planting Scheme (WTTPS), which has been parallel in timing and goals to the Community Rainforest Reforestation Program, was also established under the structural adjustment package following the cessation of rainforest logging. However, the WTTPS has a strong environmental focus to reestablish key wildlife corridors, and riparian lands, and to rehabilitate degraded lands. The Scheme has continued with Commonwealth support through the Natural Heritage Trust and Local Government support through the NQ Joint Board. The State Government has yet to contribute substantially to this $3m per annum scheme.

The NQ Joint Board was established to coordinate and integrate activities of the Local Governments within the Wet Tropics. The Board comprises elected representatives of eight Councils and is supported by a small administrative unit. The Board undertakes operational activities in all aspects of revegetation, including development of greening plans for its member shires, formulation and promotion of regional policy for vegetation management, growing seedlings, planting and maintenance of project sites, undertaking revegetation contracts and attempts to coordinate activities of the three tiers of government and the community. To support its operational activities and regional development the Board also undertakes a brokering role in a range of labour market programmes.

In recognition of the diversity of players and goals relating to revegetation and farm forestry, the NQ Joint Board has attempted to coordinate efforts through the Tropical Queensland Vegetation Management Advisory Committee. This Committee was established in 1997 to coordinate vegetation management activities across North Queensland including the Wet Tropics and Cape York regions. Members include State agencies (DNR, DPI-Forestry, Environmental Protection Agency), Commonwealth (Environment Australia, Department of Primary Industries and Energy), statutory authorities (Wet Tropics Management Authority), local government (NQ Joint Board), community (Greening Australia Queensland, Australian Forest Growers) and research groups (Cooperative Research Centre for Tropical Rainforest Ecology and Management). The Committee's goal is to ensure integration of activities to achieve production and conservation benefits. A Regional Plantation Committee with broader industry representation is currently being formed to operate alongside this Committee with a strong focus on industrial forestry issues across the region.

North Queensland is a unique case in which to examine impediments to farm forestry. It has had a high level of agency support, albeit somewhat diluted, fragmented and less focused with changes in Governments and agency structures, and has pursued a variety of objectives in revegetation as well as production of timber. Unlike in other regions, farm forestry support programs in North Queensland have concentrated on small woodlots as an adjunct to normal farming operations, and placed emphasis on high-value species such as cabinet hardwoods for which silvicultural management is essential for the production of a marketable product.

18.3 EVALUATING FARM FORESTRY PROGRAMMES

State forestry services in Australia have often provided support to encourage the uptake of farm forestry. Table 18.2 outlines some of the support services that have been available throughout Australia and provides brief comment on their success.

Farm forestry and tree growing support programmes in North Queensland have proved quite successful in generating interest in planting trees on freehold land. The Community Rainforest Reforestation Program has attracted over 560 participants and reforested over 2,000ha of mostly degraded lands. Features of the Community Rainforest Reforestation Program model which have proved successful include:

- cooperation between the three tiers of government;
- an extension support programme that has encouraged landholders to accept farm forestry as a viable land use;
- provision of training in silvicultural practices required for good stand management creating greater community involvement in farm forestry;
- multiple objectives of the programme catering for many different landholder needs in tree planting; and
- incorporation of an on-going research programme including trials on species performance, site preparation, planting design (Creighton and Sexton, 1995).

The Wet Tropics Tree Planting Scheme has rehabilitated a further 500ha of degraded areas such as riparian lands and key wildlife corridors largely unsuitable for timber production.

18.4 DEVELOPING A SUSTAINABLE FARM FORESTRY INDUSTRY

There are a number of impediments to success of farm forestry programmes in North Queensland. Many of these impediments result from the institutional structure within which farm forestry operates and are therefore relevant to farm forestry programmes in other parts of Queensland and Australia.

Table 18.2 Farm forestry programmes across Australia

Farm forestry support programme	Management agency	Success
New South Wales		
Joint venture programme for eucalypt pole plantations	State Forestry NSW (State)	Some, established 1,300ha in first year
Landholder incentives for establishing woodlots and agroforestry on cleared lands	CALM (State)	Discontinued after 1 year
Extension promoting trees on farms	Agriculture NSW (State)	Not evaluated
Northern Territory		
Extension services	Greening Australia	Limited to date
Queensland		
Provision of subsidized seedlings (Tree Assistance Scheme)	DPI/DNR (State) DPI/DNR (State)	Success in establishing but poorly managed
Full range of support services in North Qld incl. subsidized establishment, extension, research, employment, marketing	plus DPIE (Cwth.) DPI-Forestry (State)	Farmers fully supporting scheme, undertaking silviculture and establishing co-operative
Joint venture program for 10ha plus monocultures		Commenced only 1997
South Australia		
Private Forestry Encouragement Scheme leasing freehold land	PISA (State)	Failed and discontinued
Subsidized seedlings, extension services,marketing support	PISA/Forestry SA (State)	Not known
Tasmania		
Incentives to landholders through grant and loan schemes	Pte. Forests Tasmania (State)	Significant up-take of private forestry in Tasmania
Industry joint venture schemes	Industry	
Victoria		
Extension programme	DNRE (State)	Significant interest in farm forestry in key areas
Expect to start industry joint ventures in near future	Industry	
West Australia		
Share farming	CALM (State)	High up-take of farm forestry, particularly in salinity affected areas and higher rainfall wheat/sheep belt
Extension programmes	Farm Forestry Advisory Service (State)	
Industry joint ventures	Industry	

Technical Issues

Poorly focused Research and Development: Technical issues are institutional impediments to the extent that research and development programmes that determine solutions to technical issues are predominantly supported by State and Commonwealth agencies. Current impediments in this area include:

- lack of knowledge of growing requirements of high value species (e.g. spacing, species combinations, thinning, pruning, silviculture).
- disease and pests such as Tip Moth in Red Cedar (*Toona ciliata*).
- poor seedling availability and quality.
- need for development of plantation establishment and maintenance techniques suitable for high rainfall, rapid weed growth environments.
- lack of information on growth rates as input to financial analysis.

To date, public sector forestry silvicultural research has concentrated on exotic plantation species and a few select native species such as Hoop Pine. High value cabinet hardwood species have been ignored, with the exception of harvesting from native forests. In hindsight, the emphasis on breeding and selection in the silviculture of *Pinus* species was inappropriate and farm forestry is now paying the price for this lack of appreciation of the value and quality of our native timbers.

Solutions: Australia requires well-targeted research and development to catch up with the technical needs for the industry's development. These efforts should concentrate on key species such as Southern Silky Oak, Queensland Maple (*Flindersia brayleyana*) and Red Cedar, rather than the current widely spread research effort that is attempting to solve multiple problems with limited resources. For example, determining the most appropriate silvicultural treatments under which to grow Red Cedar and reduce Tip Moth problems would be a major breakthrough and provide, by itself, the basis for highly profitable farm forestry industry in Australia.

Educational Issues

Educational issues concern two areas in particular, viz. lack of silvicultural expertise and lack of community awareness and appreciation of farm forestry as a viable on-farm activity. Again these issues are institutional impediments since government provides the training and extension services which promote the development of farm forestry.

Training: Until recently, only limited training in farm forestry has been available, reflecting the preoccupation with the public forest estate. The majority of training has focused on large-scale plantation management or multiple-use of native forest resources. Farm forestry or agroforestry training is now offered in Forestry degree courses and through adult training programmes such as the Master Tree Growers Certificate offered by Melbourne University. This change reflects the increased interest in the field of farm forestry.

Farm Forestry Culture: A lack of farm forestry 'culture' has been identified as one of the major impediments to the establishment of a viable farm forest industry (AACM, 1996a, p. 72; Bruce, 1996, p. 13). Adequate extension support services are required to ensure a positive community attitude and willingness to engage in farm forestry (Creighton and Sexton, 1995). Extension programmes have been associated with most government support to farm forestry, but have not necessarily been effective. Extension programmes need to provide more than an understanding of how to establish and manage a farm forest; they need to change attitudes so that farm forestry is viewed as a suitable, safe and profitable land-use option for land. Obstacles to the development of such a culture are particularly evident in a region that has had government remove harvesting rights to timber resources through the World Heritage listing. This has created a general fear within rural communities that timber rights on freehold lands could be swept away by an expansion of the World Heritage Area.

Solutions: The challenges grouped within this set of issues reflect the transitional nature of the current forestry industry. As farm forestry develops and the preoccupation with public sector forestry diminishes, market forces will ensure that universities and other training institutions offer courses in farm forestry silviculture. Likewise, removal of institutional impediments to security of timber products will reduce the culture of suspicion of government intention. In the short to medium term extension activities will still require government support and need to be directed particularly to technology transfer. If the programmes and extension activities deliver quality trees with a commercial value then culture will evolve.

Institutional Issues

Many impediments arise from the planning and management of farm forestry support by government agencies and legislation at federal, state and local levels. These can be grouped into six major areas (Table 18.3). Suggestions for overcoming these impediments are indicated in the table.

Economic and Market Issues

Public-sector dominated pricing structures: The strong market position of the public sector in timber production and supply of logs from public sector forests effectively sets market prices below actual cost price and creates regional selling monopolies (AACM, 1996b p.17).

Table 18.3 Key institutional issues in farm forestry

Impediment	Solution
Tree tenure ambiguous and insecurity of landholder rights to harvest trees: Until recently, no mechanism to separate ownership of land and trees; insecure harvest rights, limited plantation market and investor security (AACM, 1996b, p. 17). Absence of secure future harvest rights (QDPI, 1995, p.7). Continued controversy on forest issues (AACM, 1996b, p.17). Clearing vegetation may be classed as a restricted activity - farm forestry utilising native forest resources in a sustainable fashion could then be jeopardised.	Policies allowing the separation of tree tenure from land tenure and setting out clearly where harvest rights exist and where clearing of trees may be restricted are required.
Inconsistent and uncertain land use controls (Bruce, 1996, p.13): Local Government Planning Schemes can restrict farm forestry activities or cause future harvest rights to be insecure. Forestry not considered an 'agricultural land use' (Plantations 2020, 1997).	State Planning Policy to ensure consistency of treatment of farm forestry by local government – or at least regional policies agreed to by groups of Local Governments.
Difficulties in identifying support schemes and agencies that fund farm forestry programmes: Large number of poorly coordinated and integrated programmes targeting farm forestry confuse the investment environment (AACM, 1996b, p.3). Identification of farm forestry funding agency is difficult for the general public with restructuring which at state level (Dept Forestry -> DPI -> DNR) and lead agents in farm forestry changing (i.e. PJVS – DPI now replacing CRRP – DNR).	Leadership of one agency ; recognition that forestry is long term. Establish a Trust divorced from political and departmental vagaries, Australia-wide (e.g. managed through Australian Forest Growers), statewide, or region specific (e.g. NQ Joint Board). Trust resources could include initial grants from all tiers of Government, and levies on all regional forestry operations, including public-sector forestry in both costs and benefits. For private sector contributions, co-operatives would be essential as the lead agents.

Table 18.3 (continued)

Support not designed for long-term crop: Long-term nature of farm forestry vs. short-term agency commitment. Duration of funding support schemes not usually long enough to attain cycle for self-sufficiency. An example in Queensland is where government changes have seen the demise of the CRRP/TAS and development of totally new farm forestry support through the PJVS. Uncertainty of future funding for tree planting schemes and farm forestry extension. Programmes bound up in an annual cycle of seeking funding make tree growing including stock supply and related logistics difficult.	A long-term commitment to funding needs to be built in to support programs such as the NHT Farm Forestry Program and as above, if possible, systems to be at least partially self funding and separated from political and departmental vagaries.
No recognized standards for best management in farm forestry: Current Code of Practice for forestry (DNR) not suited to farm forestry – no standards widely recognized.	Need for a Code relevant to farm forestry plus some form of accreditation system – could be task of Trust or Co-operative or ideally be one of the support actions provided by Government.
Infrastructure not designed to support forestry: Lack of infrastructure planning by local governments (Plantations 2020, 1997).	Government assistance by recognizing community benefit of farm forestry (e.g. guaranteeing minimum price) would foster private investment in infrastructure.

Solution: Major reform is required to reduce the political intervention that currently dominates the marketplace and provides very low priced timber. The first step in this process is to undertake a resource accounting exercise on the real costs of production in State Forests. Some form of collective Australia-wide policy is required to drive this process in a similar way that the COAG Water Reform agenda has encouraged and facilitated a move by all States and Territories to full water pricing and revised administrative arrangements for water resource development and management.

Product Marketing: Assistance in developing timber utilization and marketing strategies has not eventuated as yet and growers have no secured commercial prospects from their standing and planted forests. This will be critical in North Queensland where the timber industry, plantations, freehold native forest blocks and markets are all relatively small.

Solution: There is a requirement for marketing support for the growers because market forces have not yet ensured that a market for niche rain-forest timbers develops (Creighton and Sexton, 1995). Timber Co-operative based processing and marketing using standing timber will secure these markets. This will also provide much needed short-term capital until plantation resources come on stream in about 25 years, solving the problems of lack of cash flow while plantations grow to harvest age.

Export Controls and Taxation Impacts: Current export controls act as a disincentive to exporters by creating uncertainty over access to markets and interfering with market signals (AACM, 1996b, p.17). This is compounded by taxation impediments, including a lack of effective income equalization deposits (IED) or income averaging, inconsistent treatment of forestry, inconsistent capital gains tax (CGT) provisions, unclear cost sharing frameworks and a lack of cost/value indexation (AACM, 1996b, p.17)

Solution: Conduct a major review and evaluation of forestry activities as part of the Australian tax reform process. It should be noted, however, that export controls only affect a small proportion of the timber industry. It is expected that farm forestry will predominantly supply domestic markets. Parallel concerns exist with imports. There may be requirements for import controls, particularly, without accreditation or green labelling. Rainforest timbers are currently being imported and sold at comparatively low prices due to limited requirements on neighbouring countries for ecologically sustainable logging.

Risk Issues

Farm forestry is a high-risk industry. The timber crop is subject to many risks over the long rotation period, including the possibility of fire, disease, changes to land management regulations or other legislative and institutional changes that could affect the value and marketability of the final product. The long-term nature of the crop can also make forestry a poor competitor with other land uses. With annual crops, land managers can change production to meet market demands and take advantage of high market prices. Some landholders are unwilling to undertake growing of timber crops because they lose this flexibility.

Solution: Appropriate silvilcultural management may reduce risks. However, management will only be undertaken if timber is appropriately valued or, as is the case of current State Forest management, real costs of management are cross subsidized by Government.

North Queensland Specific Issues

Additional obstacles or issues with specific importance in North Queensland. There has been confusion between farm forestry support agencies with two farm forestry schemes (CRRP managed by DNR and Joint Venture Scheme managed by DPI-Forestry). The amount of technical and R&D input into the development of silviculture systems for mixed species plantations and rainforest species in plantations in high rainfall areas has been limited. Difficulty has arisen in devising marketing strategies for small volumes of a diverse variety of species, which are more difficult to coordinate than marketing of large volume of consistent product such as that produced with *Pinus* species. Conservation pressures are high for rainforest resources, particularly within the Wet Tropics region that is recognized for its outstanding conservation values; this leads to the potential for community conflict in the sustainable logging of existing private rainforest resources.

18.5 CONCLUDING COMMENTS

The outcomes achieved by the Community Rainforest Reforestation Program demonstrate that assistance programmes can be successful in promoting the uptake of farm forestry. The key to building on this success is providing a secure, long-term support programme that is not reliant on political cycles and departmental vagaries. One avenue for achieving this would be the establishment of some form of Trust to control and direct the development of the industry. This needs to be complemented by supporting and facilitating farmer-based co-operatives that coordinate and assist growers in silviculture, processing and marketing of farm forestry products. Additional to these two key institutional requirements, activities needed to foster the success of farm forestry include:

- working with local government to incorporate secure rights to farm forestry activities within local planning schemes;
- ensuring that extension programmes continue and are adequately resourced to support local farm foresters, promote farm forestry to the wider community and support farmer based cooperatives;
- developing marketing strategies for high value products;
- reforming the marketing structures for timber in Australia; and
- focusing research and development on high value timbers.

REFERENCES

AACM (1996a), *Commercial Farm Forestry in Australia – Development of a Strategy Framework– A Resource Book*, Canberra: RIRDC/LWRRDC/FWRDC.

AACM (1996b), *Commercial Farm Forestry in Australia – Development of a Strategy Framework – A Synopsis*, Canberra: RIRDC/LWRRDC/FWRDC.

Bruce, S. (1996), *Planning Impediments to Farm Forestry in North Queensland*, report prepared as part of course requirements for completion of the Postgraduate Diploma in Tropical Urban and Regional Planning, Townsville: James Cook University.

Commonwealth of Australia (1992), *National Forest Policy Statement: a New Focus for Australia's Forests*, Canberra.

Creighton, C. and Sexton, G. (1995), 'The Queensland Community Rainforest Program – lessons learnt and recommended actions to implement farm forestry', published in *Proceedings, Australian Forest Growers Annual Conference*, Victoria.

Cummine, A. (1995), 'Farm forestry – prospects, problems and policies', *Forest Resources 95 Conference*, Sydney: Macquarie University.

Forests Taskforce (1995), *Wood and Paper Industry Strategy*, Canberra: Commonwealth of Australia.

Plantations 2020 (Plantation 2020 Vision Implementation Committee) (1997), *Plantations for Australia: the 2020 Vision,* Ministerial Council on Forestry, Fisheries and Aquaculture, Canberra: Standing Committee on Forestry, Plantations Australia, Australian Forest Growers and National Association of Forest Industries.

QDNR (Queensland Department of Natural Resources) (1998), *Community Rainforest Reforestation Program Database*, unpublished, DNR Atherton (initiated by the. Queensland Department of Primary Industries in 1995).

Robins, L., McIntyre, K. and Woodhill, J. (1996), *Farm Forestry in Australia: Integrating Commercial and Conservation Benefits*, Canberra: Greening Australia Ltd.

19. Stumpage Price Determination and Investment in Australian Forests: Implications for Plantation and Small-scale Producers

M.J. Quayle and M.A. Cox

This chapter addresses the issue of log pricing methodologies before and after major reforms in the Queensland timber industry in the early 1990s. It assesses the impact of industry structure on price determination at a State level, focusing on supply and demand characteristics in the log market. It is concluded that the current bilateral monopoly arrangements deliver inefficient market outcomes with profit shares dependent on market power, but support capital investment and size economies. Location, volume and transport factors disadvantage farm foresters leaving them as price-takers with less security of gaining guaranteed sales.

19.1 INTRODUCTION

State forest agencies in Australia were for many years constrained in their management of forest resources by the requirement to meet multiple objectives. These objectives included the development of local timber processing industries, maintaining levels of rural employment, conserving the local environment and adding to improvements in the balance of payments by replacing imported timber with domestic supplies. Prices were set and administered to achieve these objectives rather than using the market to establish the value of the resource. In the 1990s, public sector reform in the timber industry saw most state forest agencies released from these ministerial directives and encouraged to pursue more commercial practices. Corporatisation, wider marketing powers and international certification standards were introduced. However, state agencies still have control and ownership over most native forests and government plantations while large processors are still located near these plantations giving them a transport cost advantage over competitors. This form of market structure leads to prices that are not competitively determined and that fail to distribute the efficiency benefits equitably between buyer and seller.

This chapter examines market structure and its influence in price determination in the Australian timber industry. Economic efficiency and competitive pricing and supply and demand conditions in the timber industry in the 1990s are discussed, and the nature and workings of a bilateral monopoly market structure that operates in Queensland. Implications of market structure for small-scale producers is examined.

19.2 ECONOMIC EFFICIENCY AND TIMBER PRICING

Efficiency in resource allocation within an industry and between industries requires that in the short run, prices for resources are determined through a competitive market environment, and these prices signal optimal supply and investment in the longer term. The concepts of optimal prices and optimal investment are therefore closely linked. Competitive pricing should assure efficiency in resource allocation in forest industries.

Forest outputs are diverse and complex, being influenced by both biological and market conditions. A variety of timber species, different end-market uses and various ownership and control arrangements all affect how forest resources are used. In Australia, hardwoods are logged mainly from state controlled native forests with eucalypt being the dominant genera. Exotic species are grown along the east coast of Australia with radiata pine (*Pinus radiata*) mainly in the southern states, and slash pine (*Pinus ellioti*), caribbean pine (*Pinus caribaea)* and hybrids being grown in northern Australian. Using this classification, there are a large number of markets based on species and location.

Social efficiency requires appropriate prices be paid by timber mills and other processors for logs from plantations and native forests. Optimal pricing of a renewable biological resource implies that the price of the resource as an input to industry be internationally competitive, and that the resource use be ecologically sustainable. Economic theory indicates that the efficient resource allocation takes place where the marginal production cost equates with value of the marginal product to the consumer. It is this rationale that promotes competitive markets as the vehicle for achieving economic efficiency and optimal resource allocation in the short and longer term. In practice many distortions and impediments to achieving efficient prices and competitive market outcomes exist, leading to misallocation of society's resources and net losses in social welfare.

Various authors have argued that state forest agencies in the past set prices below that which would operate in a competitive market.[1] In particular, Montgomery (1987) argued that such log prices would generally act as ceiling prices for logs supplied by owners of small plantations who, because of their small log output, have little market power. It is believed that under-pricing of logs by state forest agencies has reduced forest profitability and been an impediment to entry of small plantation growers.

Many factors can influence price determination of unprocessed logs. Structural and supply constraints in forest production, ability to pay by buyers, and the nexus between buyer and seller to secure continuous supply and guaranteed sales are important in log price determination. High concentration of industry structure for both growers and processors is a major determinant in price setting agreements; vertical integration, mill size, technology standards of processing plants, and intensity of resource utilisation all affect willingness to pay for timber by the processors. Other factors (many of which are socio-political in nature) impede efficiency in production and sales, including government controls designed to ensure value-adding in processing industries, the development and the promotion of the processing industry, and employment objectives.

19.3 SUPPLY CONDITIONS IN THE AUSTRALIAN LOG MARKET

Ecological, technical and economic constraints on the supply side of the log market are likely to affect stumpage prices. Supply factors such as ownership of the resource, monopoly control, government regulation, tenure of access rights for major processors and conservation requirements all have the potential to influence market behaviour. Each of these will be considered to determine its impact on price setting.

Resource Ownership

Currently, Australia's plantation resources stands at just over 1.1 M ha, 80% of which are softwoods (Burns *et al.*, 1997). Radiata pine (*P. radiata*) accounts for 73% or 883,980ha of softwood plantation area, while the native hoop pine *(Araucaria cunninghamii*) and caribbean pine (*P. caribaea*) account for a further 11%. Hardwoods (mainly eucalypts) account

---

[1] See Watt and Douglas (1974), Byron and Douglas (1981), Senate Standing Committee on Trade and Commerce (1981), DASETT (1988), CIE (1990), O'Reagan and Bhati (1991) and Resource Assessment Commission (1991).

for 157,000ha of plantations. The growth in both public and private holdings of plantations between 1987 and 1995 is indicated in Table 19.1. Softwood plantation area has increased only moderately over this period, but hardwood plantation area has more than doubled.

Table 19.1 Australian plantation ownership, private and public, 1987–95

State	Forest area, 1987 (ha)	Forest area, 1995 (ha)
Softwood plantations		
New South Wales	248,180	237,616
Victoria	204,295	196,029
Queensland	166,995	168,695
South Australia	95,317	101,907
Western Australia	87,750	88,803
Tasmania	68,000	71,972
Northern Territory	0	4,225
ACT	16,459	14,596
Sub-total	876,626	883,843
Hardwood plantations		
New South Wales	27,220	32,636
Victoria	17,872	19,347
Queensland	0	1,295
Western Australia	963	42,042
Tasmania	23,000	62,018
Northern Territory	0	0
ACT	95	263
Sub-total	83,990	157,328
Total	885,016	1,042,171

Source: ABARE (1998).

State forest services control almost 70% of Australia's plantation resource, with governments having control over planting, sales, harvesting and quota allocation. Some states have large tracts of commercial plantations of both softwoods and hardwoods on crown land. Softwood plantation timber output is increasing, with radiata pine now challenging many traditional hardwood timbers and imported timbers as the major source of supply to building and construction industries. Of the 19.1 M m^3 of logs harvested in 1996–97, 51% were softwoods (Burns *et al.*, 1997).

Long-term Contracts of Log Sales in Australia

A high level of uncertainty arises with respect to returns from forestry investments. Plantations take between 25 and 40 years to yield a high volume timber crop. Uncertainty about expected demand in the future, new technologies leading to faster growing species, decreasing management costs for maintaining the forest, changes in the interest rate, and increasing substitutes all affect potential returns from forest estate investment. It is sometimes argued that government agencies are in a better position to carry the risk of future returns than the private sector. For example, Krutilla and Eckstein (1958) note that the insurance principle of pooling greatly reduces the relative dispersion of poorly performing and highly performing projects for the government programme as a whole. It is argued that smaller private operators have neither the resources nor the financial capital to carry the risk of long-term forest investment. This may be the very reason why state governments invested in plantation forests and have such a dominant control of the timber supply and influence over log prices.

In the past, Deeds of Agreement between state forest agencies and processors for periods of five to 20 years were common in Australia. Table 19.2 (based on Margules *et al.*, 1994) compares length of contract, method of allocating log quotas, log price setting method and targeted rate of return to the forest authority, for Australian states. In recent years, the trend by state government agencies has been to reduce contract period to three to five years. With most state forest agencies operating a policy of non-transferability (although Victoria introduced transfers in 1987) and assigning new harvesting rights to encumbent processors, new entrants are restricted from the log market.

Monopoly Suppliers in the Log Market

State forest agencies along with a small number of large private companies dominate forest ownership and log supply, giving a near monopoly supply to these producers at regional locations. In Tasmania, 57% of plantation area is owned by a single private company North Limited, whereas in Queensland, only 7% of the plantation area is privately owned.

Table 19.2 Log allocation and pricing in Australian states

State	Term of contract	Method of allocation	Log pricing methods	Forest authority objectives
NSW	Up to 15 years	Contracts are renegotiated at their completion (any additional wood is put for market tender	One to one negotiation held at six monthly intervals	The target for return on assets of 7% by 1995
Victoria Victorian Plantation Corpn. (VPC)	Long term 40 years by govt. legislation; 15 years by Govt.-Industry policy; spot sales for special case supply	Agreements are formalised by a Memorandum of Understanding or an Act of Parliament. VPC allows contract changes by mutual agreement and competitive bidding for a licence for long-term access to a resource	Prices are determined by Royalty Equation System (RES) which ensures equity in pricing. Annual review adjusts prices considering relevant economic and market indicators. A review in 3 to 5 years re-establishes the base price	Target 6% return on assets. Market value sales philosophy. Tradable access licence allows competitive sales of access rights on shorter term
Queens-land	10 to 20 years	Expressions of interest for wood as it becomes available	One to one negotiations held annually, prices based Consumer Price Index or Wood Weighted Index (WWI) adjuster. Major price reviews every five years	Minimum 3% rate of return on assets is target rate. (Recently rescheduled at 7%)
South Australia	10 to 20 years	Majority of timber sold to department's processing arm	Annual 1 to 1 negotiations, effective from 1st July. Prices based mainly on sawn timber price movements and profitability of department's processing arm	To earn an acceptable rate of return on assets
Western Australia	5 to 15 years	Established purchasers given first option on renewal. About 10% of sales are by open tender	Non-negotiable target royalties	To maximise profits
Tasmania	10 years; renegotiated every 5 years	Current negotiated contracts retained, but all maturing parcels of timber put up for tender	Two thirds by negotiation, one third by tender	6% rate of return on assets

The large tracts of public-owned plantations are the result of a number of Commonwealth Government forestry Acts between 1967 and 1978 that set targets to increase plantation area and which focused on current suppliers expanding their areas of conifers. This has placed control of log supply and price setting in the hands of a few large growers (especially the state governments) in each state. Little emphasis was placed on encouraging and diversifying the ownership of plantations or promoting private small-scale forestry initiatives similar to the Small Business Forest Enterprise programme in Canada. Initiatives of this kind only emerged in the late 1980s and early 1990s as part of the farm forestry programme to encourage a diversification of ownership and an expansion of the plantation estate. Many small-scale foresters plant more for environmental reasons than commercial gain, and pose no competitive threat to the large private companies or state forest agencies.

Large growers are in the position to set and negotiate the terms and conditions of supply to timber buyers. Small privately owned plantations are generally located further from the mill and suffer location, volume and transport disadvantages in influencing log prices (Quayle, 2000). As regional monopolist suppliers, state forest agencies can negotiate prices in excess of the opportunity cost of the timber and gain some of the monopoly rents. Smaller growers will survive with normal profits if they locate near large processing plants that require additional timber supplies as they expand their throughput. State forest agencies are now using their market position to strike higher prices. Farm foresters have little or no market power and have to accept prices processors are prepared to offer. Hence most small-scale foresters are price-takers and marginal suppliers, accepting prices below the royalty rates set by the large growers.

New Directions for State Forest Agencies

With major reforms in the public sector in the early 1990s, many of these state forest agencies were directed to pursue new objectives. In Victoria, the formation of the Victorian Plantations Corporation (VPC) introduced the transferability of long-term contracts and the sale of parts of the forest estate to private interests. Recommendations were made to the VPC that criteria for stumpage determination include recovery of full market value of standing timber, ensuring a close relationship between log prices and end-product prices and thereby encouraging future investment in forestry. Other states established commercial business groups within government departments, with responsibilities that included the management and marketing of state-owned plantations, the promotion of joint ventures, and progress to more market-oriented outcomes. In Queensland, the business goal of DPI Forestry was clearly stated as 'maximising the return on its assets . . . and to pay to its shareholder, the State government, an annual dividend on revenue earned' (DPI-Forestry 1998, p. 23).

State forest services remain subject to strong political lobbying from sawmillers and timber processors to maintain lower administered log prices. As a result, these newly commercialised forest services are still constrained in their marketing and price setting strategies, which could well prevent them achieving maximum returns on their forest assets.

19.4 THE NATURE OF DEMAND IN THE AUSTRALIAN LOG MARKET

Timber demand and prices are influenced by a number of factors. Prices that processors are prepared to pay for logs depend on the structure of the industry, the size and efficiency of processing plants and other factors such as prices received for processed timber in end-markets. The timber processing industry is both fragmented and unorganised. At first glance it appears to be a highly competitive industry with many plants and firms. In 1988–89, there were about 1,160 hardwood and 240 softwood mills in Australia. Little rationalisation in industry structure has occurred in the past decade. In 1998, there were still about 1,200 mills, with about 10% of them producing about 90% of the sawn timber (BRS 1998, p. 127). Mills are characterised as either 'fixed-location' or 'portable'. The majority of mills (81%) concentrate on native forest timber. As DPI-Forestry (1998, p. 39) noted, 'although all fixed location mills transform log inputs into salable products through a sawing process, there is a clear difference in the nature of those mills processing native resources and those based on plantation timber.' The different nature of these mill types means that they can almost be considered as two separate industries.

Many of the fixed-location and portable mills are small in size with low throughput. Approximately 94% of the hardwood mills and 80% of the softwood mills have log intakes of not more than 15,000 m^3/year. Margules *et al.* (1994) suggested that in the Australian timber commodity market, softwood sawmills with annual log inputs of 200,000m^3 or more are internationally competitive in terms of wood processing costs. Their processing costs are similar to those in New Zealand, less than those of British Columbia and the Pacific North West of the USA, and slightly more than in Chile (which has lower labour costs). In Australia, only the large dedicated capital intensive processing plants satisfy this size criterion. Industrial companies in Australia own over 65% of the private plantation estate (Hogan *et al.*, 1997), most being large vertically integrated firms, e.g. Australian Paper, Australian Newsprint Mills and North Forest Products. These companies tend to concentrate in specific regions and specialise in growing plantations for sawlogs, medium density fibreboard or paper products.

Firms specialising in commercial softwood plantation timber generally have large plants, are capital intensive and have a considerable amount of 'high-tech' equipment. They tend to be highly localised and dominate the region for access to timber resources. In some cases they own the resource or have long-term concessions from state forest agencies. In Queensland, plantation mills make up only about 12% of the total number of fixed-site sawmills but account for 60% of the overall log input of 1.6 M m^3.

19.5 BILATERAL MONOPOLIES IN THE AUSTRALIAN LOG MARKET

Established timber processing firms have cost advantages over potential entrants. Even after long-term log concessions have expired, these firms are efficiently located to minimise transport costs and to increase processing capacity at lower marginal cost than new entrants. Processors act as *monopsonists* with considerable power to influence the price of logs from local plantations. While there is scope in some Australian states to move to competitive market pricing through spot market tender bids and auctions, the reality is most commercial plantation forest resources are in the hands of state forest agencies that can set prices higher than the cost of producing the timber. Similarly, large processors with scale economies and locational advantage are in a monopsonist position as a buyer for the timber resource once committed to the region (Ferguson, 1982). These processors tend not to be price-takers for timber, influencing prices considerably by tendering prices closer to the cost of timber production in an attempt to capture economic rents. Hence a *bilateral monopoly* market structure exists.

Because the value to the processor of additional timber is higher than the cost of production of the timber, the buyer is prepared to pay a price that is higher than the marginal private cost of making additional timber available to the market in the short run. Additional timber is available to the processor from the spot market, but normally the spot price will be higher than the long-term contract price. Processors will seek additional quantities of timber up to the point where the expected revenues from using the additional timber equate with the cost of procuring the timber. This situation is illustrated in Figure 19.1, which depicts the short-run supply and demand conditions in a regional log market (S_o and D_o respectively). The marginal factor cost (MFC) line represents the maximum amounts the processor is prepared to pay for log input according to their expectations of the additional revenue from the sale of the processed timber. MR is the marginal revenue per additional unit of output.

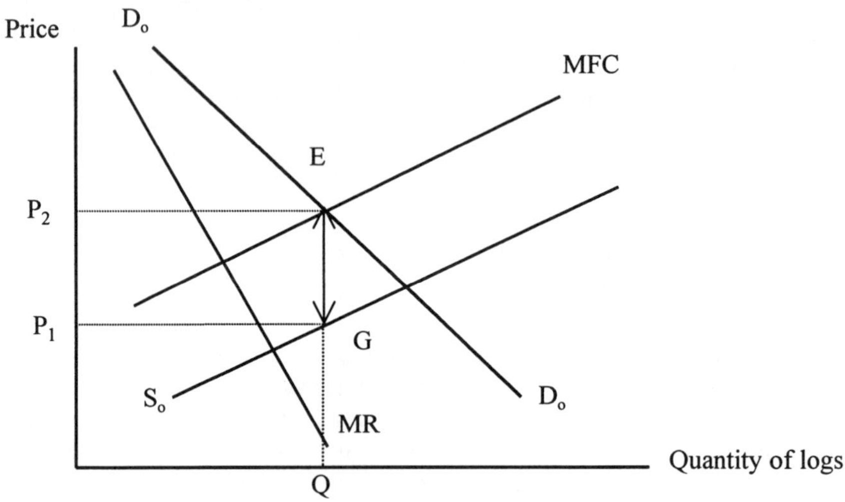

Figure 19.1 Bilateral monopoly negotiable price range in the Australian log market

At a specific quantity of timber demanded, say Q, the maximum amount that the monopsonist is prepared to pay is P_2. At point E, the buyer's additional revenue from the use of the timber equates with cost of obtaining the additional timber. If the supplier is aware that the processor is prepared to pay P_2, then this will be the price that the supplier will wish to achieve in price negotiations. On the other hand, the buyer prefers to pay as little as possible but would be prepared to pay P_1 for the quantity Q, as this will maximise the processor's profit. This outcome is shown as point G in Figure 1. The difference between P_1 and P_2 is the 'negotiable price range' in which price will be indeterminate. The grower will attempt to push the price for the resource towards P_2 in negotiations, while the processor will try to drive the price down to P_1.

Under these market conditions, efficient resource allocation yielding maximum social welfare will not be forthcoming because the 'settled' price will not be a competitive market solution where the marginal value of the resource equates with the marginal cost of producing an additional unit.

Negotiating a price for a particular volume of timber may require a stumpage appraisal system to determine the minimum reservation price or 'upset' price that the grower is prepared to accept so as to cover the costs of growing the logs.

Hyde (1980, p. 224) criticised the stumpage appraisal system as 'administratively costly and even unnecessary except where the forest service was facing colluding buyers or a monopsonist'. Moreover, Ferguson (1982, p. 245) recognised that 'forests, processing plants and major consumption centres for the final products are seldom distributed uniformly...[which] may enable particular producers or consumers to exercise a degree of monopoly or monopsony power'. These conditions still persist. Under the conditions of bilateral monopoly, competitive market outcomes will fail to be achieved and markets will fail to deliver a socially optimum allocation of resources or appropriate incentives for future investment in the timber industry.

19.6 DETERMINING LOG PRICES IN QUEENSLAND

Determining the price of timber is a complex exercise. Timber is not a homogeneous product, but differs in its characteristics with respect to hardwood and softwood, native forest and plantation, strength, size and durability across species. It is a versatile product in terms of uses, and is channeled into many markets, including sawlogs, poles, plyboard, veneer board, and laminated veneer lumber (LVL). Queensland's tropical timber species are used in many of these end-products. Pricing formulae applied by a state forestry agency should include parameters that cover diversity of species, location, end-market use and growth rates. Other factors such as accessible volume and extraction costs are likely influence the formula through discount allowances. Variations in these parameters lead to different prices for different forest resources.

Most Australian State forest agencies nowadays combine market-related pricing mechanisms in the form of auctions or tenders with one-to-one negotiations. Queensland's *propositional call* mechanism was introduced in the early 1980s and continues to be applied today. Propositional calls are a form of closed tender which are submitted after the forest agency calls for expression of interest for harvesting and access rights to a particular stand of timber. The submitted tenders form the basis for price negotiations.

All Australian states are moving to shorter contract terms and more market-related pricing processes but still include one-to-one negotiation between the grower and the processor as an important element in price setting. Queensland moved towards a variable royalty rate methodology in the early 1980s with royalties calculated according to tree characteristics and potential sawn timber recovery rates. These calculations are used to establish a threshold price. Higher quality stands of timber can expect higher bid prices. Close negotiations between buyer and seller remain a dominant feature of forest sales in Queensland, which can exclude new entrants to the market. Without the threat of entry by rival processors, competitive market bidding simply will not emerge in this environment.

Queensland has been at the forefront of reform of price determination mechanisms. In the early 1980s the then Department of Forestry replaced the ageing residual stumpage assessment mechanism with a variable royalty algorithm and replaced open auction selling with a negotiated propositional call procedure. In 1996, Queensland, through the use of a pilot program, introduced a contract price adjustment mechanism designed to reflect the current market value of timber in previously negotiated sales and the value of the processed timber in various end-uses. Contract sales are now not only of a shortened term (three to five years), but also adjusted more frequently using indexes that more closely reflect activity in end-product markets. Overall economic conditions are also taken into consideration in determining price adjustments. Other states are also implementing these price adjustment mechanisms, albeit slowly.

As under the previous residual pricing approach, the contract or 'base' price is still considerably influenced by the processor's capacity to pay. Smaller, less efficient millers and processors with older technologies and higher costs can incur up to 60% of their total input costs from log purchases. High prices for timber often can make these operations unprofitable and lower prices may be negotiated as part of the propositional call system. If negotiations result in prices lower than that which could be obtained on a more open market, then the price determination mechanism will inherently be inefficient. Low prices signal a misallocation of resources discouraging plantation investment for both large and small-scale foresters. Processors too will have little incentive to invest in more efficient, lower cost processing equipment if log input prices remain low.

Over the past decade DPI-Forestry have negotiated price increases for most of their log contracts, particularly from softwood plantations. Through the use of the consumer price index and more recently the wood weighted index as price adjusting mechanisms, softwood plantation timber prices increased more rapidly than in other states from an average of \$36.20/m^3 in 1987–88 to about \$75–\$90/m^3 currently. Some tropical log prices have been adjusted back by 25% because of increased competitive pressures from low-priced imported radiata pine.

19.7 CONCLUSIONS AND IMPLICATIONS

Major reforms were implemented by most Australian state governments in the 1990s with the objective of correcting market inefficiencies in the timber industry. The high concentration levels of market power on both the supply and demand side of the market generate a process for price setting still based on close negotiations between buyer and seller with little competition for specific blocks of timber. While the duration of contracts has been reduced and log quotas made transferable among processors, little change to the industrial structure of forestry has occurred. Market power influences log prices now even more than before these reforms were effected. Establishing base prices and adjusting prices of continuing contracts is still determined through a process of negotiation between supplier and buyer, despite timber packages being offered through more market-related mechanisms of closed tender, propositional calls and spot price offerings.

The bilateral monopoly situation ensures that both the buyer and the seller share the economic rents from the resource, the proportionate shares depending on the negotiating power of each. This outcome is less efficient than a competitive market solution, reducing overall output and incentive for investment in plantations. Despite these efficiency shortcomings, a bilateral monopoly industrial structure offers some advantages. On the demand side, it offers the processors guaranteed timber supply to sustain the capacity of their capital investments, and gives incentives for large, well-located processors to expand operations to reach critical-mass size for international competitiveness. Scale economies and technological advancement are seen as the means to compete in timber markets against countries with lower labour costs. On the supply side, these market arrangements ensure growers share in the profits earned by the resource. Location is important because transport costs are critical in the returns equation. The further the plantation from the mill, the lower the return to the resource, regardless of species type. Small private timber producers are unlikely to have competitive advantages in securing returns similar to larger, better located plantations. As processors increase their operating capacity to a size that requires continuous and additional timber supplies, owners of small forest areas will be encouraged to invest in additional forest in an attempt to capture the marginal demand from the processors, but are likely to remain price-takers.

REFERENCES

ABARE (Australian Bureau of Agriculture and Resource Economics) 1998, *Australian Forest Product Statistics*, March Quarter, Canberra: ABARE.

BRS (Bureau of Resource Sciences) (1998), *National Forest Inventory, Australia's State of the Forest Report*, Canberra: National Forest Industry.

Byron, R.N. and Douglas, J.J. (1981), *Log Pricing in Australia: Policies, Practices and Consequences*, Canberra: BFE Press.

Burns, K., Hansard, A., Lawson K. and Ball, A. (1997), 'Plantation Development in Australia: Some Considerations', *Australian Forest Product Statistics*, December Quarter, Canberra: ABARE.

CIE (Centre for International Economics) 1990, *Optimum Pricing Arrangements for Tasmania's Crown Forests*, Canberra: Centre for International Economics.

DASETT (Department of Arts, Sport and Environment, Tourism and Territories) (1988), *Report of the Commission of Inquiry into Lemonthyme and Southern Forests*, Vol.1, Canberra: Australian Government Printing Service.

DPI-Forestry (Queensland Department of Primary Industries – Forestry) (1998), *An Overview of the Queensland Forest Industry*, Information Series Q198105, Brisbane.

Ferguson, I.S. (1982), 'Selling Wood from State-Owned Forests', *Australian Forestry*, 1982, **45**(4), pp. 245–254.

Government of Victoria (1986), *Victoria: Timber Industry Strategy*, 9, Melbourne.

Hyde, W. F. (1980), *Timber Supply, Land Allocation and Economic Efficiency*, Baltimore: Johns Hopkins Press.

Hogan, L., Gilmour, L., Weston, L. and Lawson, K. (1997), 'Regional Forest Agreements – ABARE'S Economic Assessment', *Australian Forest Product Statistics*, December Quarter, Canberra: ABARE.

Krutilla, J. V. and Eckstein, O. (1958), *Multiple Purpose River Development: Studies in Applied Economic Analysis*, Baltimore: Johns Hopkins Press.

Margules, Groome and Poyry (1994), *Victorian Plantations Softwood Stumpage Review and Outline of Price Adjustment Mechanisms*, prepared for the Victorian Plantations Corporation, Canberra.

Montgomery, P. (1987), 'Private plantation pine: what price the wood?', *Investment Seminar*, Australian Forest Development Institute, Melbourne.

O'Reagan, M. and Bhati, U.N. (1991), *Pricing and Allocation of Logs in Australia*, ABARE Discussion Paper 91.7, Canberra: AGPS.

Quayle, M.J. (2000), 'Public-Private Ownership in Commercial Softwood Plantation Forestry: the effects of Industry Structure', paper presented at the *IUFRO Small-Scale Forestry Conference*, Kuranda, Queensland.

Resource Assessment Commission (1991), *Forest and Timber Inquiry Draft Report*, Volume 1, AGPS, Canberra.

Rose, R. and Bhati, U.N. (1992), Pricing of Logs from Native Forests, ABARE Discussion Paper 91.7, Canberra: AGPS.

Senate Standing Committee on Trade and Commerce (1981), *Australia's Forest and Forest Product Industries*, Canberra: AGPS.

Watt, A.J. and Douglas, J.J. (1974), Royalty Systems: Principles and Policies, paper presented at the 23rd Australian Timber Industry Stabilisation Conference, Wodonga, Victoria.

20. Towards a Vibrant Tropical Small-Scale Forestry Industry: Imperatives, Opportunities and Challenges

S.R. Harrison and J.L. Herbohn

This chapter reviews the key insights of earlier chapters into small-scale forestry in the tropics, and examines challenges and opportunities for small-scale forestry in the future. Elements in common across countries with respect to forestry in the tropics, and with respect to small-scale forestry systems, are discussed. Small-scale forestry is viewed as a distinct sector of forest industries, for which carefully targeted support policies are required. Improved information and new tree technologies should allow progressive improvement in support policies.

20.1 REVIEW OF EARLIER CHAPTERS

Preceding chapters have examined a variety of aspects of farm and community forestry in the tropics. Section I provided a background to the nature and evolution of these forestry systems. Lamb and others (Ch. 2 and 3) illustrated the evolution of small-scale forestry systems, drawing on experience in North Queensland, Australia. Landholder and community attitudes have changed radically over time, from regarding native forests as an abundant resource and an impediment to agriculture, to recognizing the need to conserve remaining forests and promote reforestation. Notably, it has been landholders who have carried out much of the experimentation in tree growing, rather than researchers in forestry agencies.

Section II has focused on the timber industry in the tropics. Improved forward estimates of plantation production and profitability are becoming available through advances in stand yield and financial modelling efforts (Ch. 4–6). In that lack of information is regarded as a major impediment to tree planting, this work offers promise for encouraging further planting. Appropriate milling technology for small-scale forestry may include portable mills, though the role for these has not been well understood (Ch.

7). Small-scale forestry is not necessarily competitive on cost grounds with industrial forestry, but offers promise for growing high quality timber species for niche markets such as furniture manufacture (Ch. 8). Even for countries which are net timber importers, scope may exist for export of specialist timbers and value-added products arising from small-scale production (Ch. 9).

The non-timber benefits of forestry are being increasingly recognized, as discussed in Section III. Chapter 10 has outlined the scope of these benefits, which can be much more important than timber production in some situations. At the same time, these non-wood benefits can be over-estimated, as revealed for example by the modest impact of reforestation on land values (Ch. 11). A particularly important form of environmental planting is riparian revegetation, a form of catchment management which can involve various stakeholder groups with diverse objectives, hence the need for a multicriteria analysis of revegetation options (Ch. 12). Scope exists for considerable improvement in discharging reporting responsibilities and providing more useful information by forest management agencies, particularly with respect to non-timber benefits (Ch. 13).

A major objective of studying small-scale forestry systems in the tropics is to provide a basis for improved public-sector policy making. In that forestry generates positive externalities beyond the benefits to landholders, there is justification of taxpayer (and aid agency) support for tree planting. In this context, Section IV examines policies to overcome impediments and encourage planting. The way in which forestry income is taxed, and allowances are made for expenditure, has the potential to influence the private profitability of forestry. Often forestry taxation arrangements are not well defined nor well understood (Ch. 14). Smallholders in developing countries in general do not pay taxation or benefit from tax concessions (and in some developed countries farm forestry is outside the taxation net).

A prerequisite for good policy making is an understanding of the attitudes and objectives of landholders towards forestry. Various landholder surveys have revealed surprisingly consistent attitudes to small-scale forestry in sub-tropical and tropical Australia, and suggest the possibility of classifying farmers into distinct groups with respect to their attitudes and objectives (Ch. 15). Experience suggests that there are a number of keys to successful small-scale forestry which appear relevant in many countries; these include including secure property rights to land and tree crops, a viable production technology, capacity for crop protection, and adequate markets (Ch. 16).

In recent years, increasing attention has been paid to community forestry programmes, important examples being the joint forest management (JFM) movement in India and community based forest management (CBFM) in the Philippines. These programmes present major opportunities for community empowerment and welfare improvement, but are also faced with a number of difficulties (Ch. 17).

Major initiatives in small-scale forestry for production and environmental purposes took place in North Queensland during the 1990s, including the Community Rainforest Reforestation Program. Chapter 18 has provided insights, from experiences in management of the North Queensland Joint Afforestation Board and other organisations, into the surprisingly long list of impediments to tree planting, and of institutional arrangements to overcome these impediments.

One of the major impediments to tree planting has been low timber prices, often arising from stumpage price policies for government plantations being based on multiple objectives including social objectives. Sales of timber from small-scale forestry often take place on the fringe of a bilateral monopoly forestry market between government and industrial timber users. Reform in stumpage pricing policies, and greater organisation by small-scale producer groups, is necessary to achieve higher stumpage prices for small-scale plantations (Ch. 19).

20.2 DISTINGUISHING FEATURES OF TROPICAL SMALL-SCALE FORESTRY

To what extent is small-scale forestry in the tropics different to traditional production forestry in temperate zones? Is it a distinct sector requiring separate policies? These questions will be examined with respect to commonality of issues, first across tropical countries and second for small-scale as distinct from industrial forestry.

Common Issues across Tropical Countries

Most nations with a considerable area of land in the tropics, and with significant areas of tropical forests, are so-called 'developing' countries, Australia being an exception. Most (including Australia) have experienced high logging rates and large-scale deforestation in recent times. As a corollary to unsustainable logging practices, most tropical countries are now experiencing a loss of an important revenue source and of the environmental services provided by forests. Illegal logging has sometimes been a problem, with control beyond the resources of government. Government and community have gradually become aware of the severe resource loss and environmental problems caused by deforestation and, with changes in attitudes and sometimes pressure from international agencies and loan and aid organisations, most are now taking steps to protect residual native forests, and to promote tree reforestation (including small-scale forestry).

Tropical regions throughout the world have some native timber species with outstanding timber qualities, which have been prized by timber industries. There is often a history of woodcrafts and furniture manufacturing. The high quality timbers have been popular in overseas markets, which has been one of the causes of rapid timber extraction.

Plantations typically have rapid growth in the high rainfall and winter-free environment of the tropics, and rotations of eight to 15 years are not uncommon (Ch. 6). Some very high quality hardwood timbers are well suited to plantations in the tropics, and are grown widely, such as the well-known species of mahogany and teak, as well as a large number of less well known and researched species. There is less interest in conifer plantations than in countries in temperate regions, and greater interest in forest products other than timber, including 'minor forest products' and fuelwood.

While rapid growth and high-value species would suggest strong interest in forestry, developing countries in the tropics are often typically poor, and face high demand on fertile land for food production, severe shortage of investment capital and high interest rates, which discourage long-term investments. Often aid and loan finance is required to undertake forestry.

Community forestry as distinct from farm forestry is of growing importance in developing countries, as a means by which smallholders can pool skills and resources. In contrast, community forestry in developed countries tends to mean community organisation rather than common property resource management. Another Western version of community forestry is the forestry co-operative, which is mainly concerned with increasing the market power of tree growers with respect to input purchases and timber sales.

The level of financial support from domestic governments for small-scale forestry is often low in the tropics compared with more affluent countries in temperate regions. However, foreign governments and NGOs often support community forestry programmes. In recent years, there has been a tendency for governments to increasingly support small-scale rather than industrial forestry, because of the social benefits it can create, and as a measure to counter illegal logging of protected forests.

Forestry plantations have not always been successful in tropical areas. For example, establishment rates are sometimes low, particularly when maintenance programmes are not well supported. Also, a high rate of loss of plantations due to wildfire has taken place in some countries, notably Indonesia and the Philippines.

Common Features of Small-scale Forestry across Countries

When contrasted with industrial forestry, small-scale forestry across various countries tends to share a number of features. Small-scale planting often means very small woodlots or field border, roadside and canalside plantings. Lack of economies of scale can mean higher establishment and maintenance costs. Small-scale growers in general lack silvicultural skills, and other resources for forestry, hence adopting low-intensity forest management which can involve little thinning, pruning or fertilizing. Producers lack market power and marketing skills, leading to low timber prices, and much of the timber is used for on-farm purposes rather than sale. Simple technology is often used for harvesting and milling, e.g. chainsawmilling, portable sawmills (Ch. 7) and small bench saws. With the short rotations, markets are geared to tolerate small-diameter immature timber.

Current and potential growers typically lack knowledge about likely financial returns from forestry. Along with the different species choice and silvicultural methods, prediction of stand yields for small-scale forestry is more difficult than for industrial plantings. Sparse data may necessitate use of different statistical approaches, and application of data from other countries (Ch. 6).

Some of the technology and practices desirable for industrial and government forestry will be of less importance for small-scale forestry. For example, there is likely to be less interest in labour saving production systems, fertilizer strategies and quality assurance with respect to pruning. There is less need for certification of sustainability, since export markets are not targeted, although it has been argued that sustainable management and harvesting fit well with the objectives of farm foresters (Ch. 9).

While government support for small-scale forestry is common, the types of support measures appear to vary greatly between countries, and to be subject to frequent change. There tends to be an emphasis on simplistic support measures such as free or subsidised seedlings, and not on tackling the difficult property rights and marketing issues which are often a greater impediment to tree farming.

In summary, small-scale forestry is not simply a scaled down version of industrial forestry, but is a distinct type of enterprise which often differs with regard to objectives of the grower, species choice, planting layout, silvicultural management, harvesting and marketing systems and other factors. Different impediments arise, and different support measures are needed.

Sustainability of Tropical Small-scale Forestry

As discussed in Harrison *et al.* (2000), generally small-scale forestry can be viewed positively with respect to sustainability. This is typically linked with the multiple goals for forestry of small-scale growers, including environmental goals, and with the environmental motives of government support programmes. Also, by its nature, small-scale forestry involves relatively large numbers of participants, and may include low-income farmers, tribals, and so on, hence having favourable social impacts. Economic sustainability is perhaps less certain, particularly if judged by income earned from timber sales. However, on-farm use of timber and non-wood benefits may increase the payoff from a wider economic perspective.

20.3 IMPERATIVES, OPPORTUNITIES AND CHALLENGES

Earlier chapters reveal both opportunities and challenges to the expansion of small-scale forestry, some of which are now reviewed.

Imperatives Which Will Drive Small-scale Forestry

Replacement timber source following excessive deforestation. One of the most compelling reasons for believing that small-scale forestry will prosper is the critical need for greater reforestation activity than is taking place through industrial forestry. Various tropical countries were major timber exporters, but now have domestic timber shortages, a prime example being the Philippines (Venn *et al.*, 2001). Environmental benefits of forestry, the need to reduce logging pressure on native forests (and on felling coconut trees for lumber), meeting domestic timber and fuelwood needs, and reducing expenditure on imports will all provide an impetus for government to support a timber industry.

Need for watershed protection. A serious consequence of deforestation in uplands is land degradation and lowland flooding, with loss of human life, crops and stock, and damage to infrastructure such as housing, roads, bridges and hydropower generators. This can be expected to provide further impetus for governments to promote reforestation. Agroforestry, community based forest management, plantation forestry, and management of native forests are all viewed as elements of watershed management (PCARRD *et al.*, 1999).

Social imperatives. Tree growing, especially of multiple-purpose species, has the potential to improve the welfare of the rural poor, including people migrating from heavily populated areas to the uplands for new livelihood opportunities as well as tribal communities. This has been the rationale for the certificate of stewardship contract in the Philippines (Aggangan, 2001).

Wildlife protection and habitat. Surveys indicate that in developed countries, wildlife habitat and landscape amenity are highly ranked reasons for tree planting. An important type of environmental planting will be development of wildlife refuges and corridors. In developing countries, as economic progress is made, or as ecotourism is pursued, more environmental plantings can be expected to occur. Also, there is now interest in new management methods for buffer areas around national parks to conserve wildlife and vegetation (Dart *et al.*, 2001).

High woodfuel demand. Continued high demand for biofuel can be expected in low-income countries, for industrial, restaurant and domestic use.

Meeting greenhouse gas reduction targets. It seems probable that countries will have great difficulty meeting the greenhouse gas emission reduction targets agreed at the Kyoto conference (Lamb, 2000). Forestry as a carbon sink can play a role in a greenhouse gas reduction strategy.

Timber as a renewable resource. Timber can replace other products, such as fossil fuels and concrete poles, which rely on non-renewable resources.

Opportunities for Small-scale Forestry

A number of circumstances have arisen which create opportunities of expansion of small-scale forestry.

Improved tree production technologies. Greater understanding of, improved site-species matching, fertilizer regimes, and thinning and pruning strategies, and their introduction, can lead to higher yields and profitability of small-scale forestry.

Introduction of new seedling technology. Improved seedling technology, as is being introduced in the Philippines (Dart *et al.*, 2001) can lead to lower-cost and higher quality seedlings for small-scale foresters.

Improved genetic stock. Tree breeding work, which has in the past been neglected for species not widely grown commercially, offers promise for more rapid growth, better tree form and greater economic viability of a wider range of tree species.

Introduction of new tree species. Often newly introduced species will outperform traditional species. In this context, Australian eucalypts and acacias have been found to perform well in various tropical countries, and tolerate a range of site conditions, including drier areas and soils of marginal fertility or acidity, and appear to have potential for farm and community forestry and agroforestry systems (Harrison and Herbohn, 2001).

Improved wood processing techniques. Techniques such as finger jointing which allow smaller logs and shorter lengths of timber to be used, and curing and new wood panelling products which add value to timber, also have the potential to increase profitability of small-scale forestry and hence encourage greater adoption.

Changing land tenure regimes. With agrarian reform and progressive granting of indigenous land tenure, there will be increasing land scarcity for industrial forestry but greater opportunity for community and farm forestry. This trend is not restricted to developing countries. For example, there is considerable interest in utilizing forest resources and perhaps reforestation by indigenous communities which are gaining increased 'native title' land rights in Cape York in northern Australia.

Carbon sequestration and tradable credits. Through carbon sequestration, forestry can make an important contribution to slowing greenhouse gas accumulation and global warming. Markets for trading in carbon credits can be expected to emerge in the near future, providing an additional return to growers. Also, use of logging residues has potential for small-scale electricity generation as a greenhouse gas neutral source of electricity.

Willingness of landholders to undertake experimentation. As pointed out by Lamb *et al.* (Ch. 2), small-scale foresters are often prepared to undertake experimentation on species selection and plantation establishment and management. Much of the knowledge in small-scale forestry in Australia has been generated because of the personal interest of inquisitive individuals through informal experimentation on their own land, in part motivated by the lack of public-sector funded research into small-scale forestry. In low-income countries, this experimentation is both at a farm level and in terms of community based forest management. Here, research is often supported by foreign aid, including assistance from non-government organisations. Community forestry systems are to some extent still a social experiment, and can be expected to be refined over time as more experience is gained.

Catch-up of the research sector. As governments, including research agencies, come to recognize the importance of small-scale forestry and the constituency of support, they can be expected to devote more effort to this area. Many foresters have been trained in a rather narrow paradigm of production forestry with monocultures of exotic conifers in temperate climates. We have experienced remarkable negativity in seeking research grants when proposals have been reviewed by these traditional foresters. In Australia there are however a number of progressive funding and research organisations that take a much broader view of forestry and that have supported our research on small-scale forestry (e.g. RIRDC, ACIAR, Rainforest CRC and NQJB)

Accumulation of silvicultural knowledge. Over time there will be an accumulation of silvicultural knowledge, e.g. about best species and mixtures, establishment methods and field layouts. Much of this will be generated by practitioners, not government. At the same time, researchers will develop improved estimates of non-timber benefits of small-scale forestry, and hence of the tradeoffs between production and social and environmental benefits, to assist in designing forestry support programs.

Improved understanding of characteristics of tree growers. A better understanding is being gained of the impediments to small-scale forestry, and of landholder attitudes and objectives in relation to forestry. This provides a sound basis for targeting of assistance measures by various levels of government, and by aid agencies including NGOs.

Greater value-adding to harvested timber. Some tree grower groups, including community forestry groups, are now investigating the training and skills required for making joinery products, furniture and souvenirs, with a view to adding value to their products and increasing the economic payoff per hectare of forest.

Improved communication systems and electrification. Demand for poles for electricity, telephones and cable television is an important and growing market, both for new services and replacements for existing infrastructure.

Challenges Facing Small-scale Forestry

It would be foolhardy to predict rapid expansion in small-scale forestry in the tropics. Even with appropriate government policies to overcome impediments, progress will be limited by a number of factors.

Long payback period and high investment risk. Forestry is a long-term and moderately risky investment, which is not feasible for many low-income farmers.

Lack of market power of small-scale growers. Evidence suggests that the stumpage prices for small-scale producers are lower than for industrial foresters, limiting financial returns.

Negative environmental externalities. Forests cannot be said unequivocally to reduce land degradation and soil loss. Compared with pasture cover, the soil surface under trees is often exposed to rain action (especially with deciduous species), and soil disturbance occurs at planting and harvest. While providing improved habitat for native wildlife, some of which reduce crop protection costs, forests can also provide habitat for economic pests. For example, in North Queensland, Australia, there is concern that farm plantings add to the habitat already existing in national parks for exotic animals (feral pigs, foxes) and native wildlife (wallabies, cockatoos) which damage sugar cane and other agricultural crops.

Difficulty of access to markets in carbon credits. Markets for carbon credits in forestry still have some hurdles to overcome before they are available to farm foresters. Even then, it will be difficult for small-scale foresters to participate in markets for carbon credits, given the transactions costs in establishing and verifying carbon sequestration amounts (Lamb, 2000). These difficulties may possibly be overcome by formation of grower co-operatives or through market intermediaries.

Limited public-sector support. Particularly in developing countries, the level of assistance to tree growers is constrained by the amount of funds governments can devote to promotion and support of forestry, and the amount which can be obtained in aid and loan funds.

20.4 POLICIES TO ENCOURAGE TROPICAL SMALL-SCALE FORESTRY

Various measures may be adopted by governments to assist in overcoming impediments and making small-scale forestry more attractive to individual landholders and community groups.

Recognize the positive externalities. Government policies to support small-scale forestry need to recognize the positive externalities from timber production, and also the important non-timber benefits generated, and be designed to maximize these benefits.

Targeting of tree grower assistance programmes. Policies to support small-scale forestry need to be designed with a clear understanding of the objectives and likely reactions of targeted beneficiaries. Direct financial assistance to tree growers is superficially an effective way to 'get trees in the ground'. However, experience indicates that this type of support, while useful for demonstration purposes, tends not to stimulate continued planting.

Removal of impediments. Insecure property rights (e.g. with respect to land tenure and harvest and transport rights) and sovereign risk can be major impediments to tree growing. These impediments are not easy to overcome, but their removal is likely to be more effective in promoting small-scale forestry than simple handout programmes such as provision of free seedlings.

Utilize the people. It has been found in India that governments have great difficulty in preventing illegal logging by monitoring methods, hence it is more effective to promote community forestry in which the community have an incentive to protect *their* forests, by allowing a stake in ownership and management (Ch. 17).

Provision of information. Extension programmes can be a relatively low-cost and effective means of bringing about increased rates of tree planting. Extension needs to be backed up with sound research, for example into suitable species to grow, desirable silvicultural techniques and likely profitability.

Work with foreign agencies. Particularly in developing countries, government policy can be supportive of and attract foreign research and financial assistance for small-scale forestry.

Promotion of good forest practice. Government, through extension and training, is in a position to encourage high standard silvicultural practices, e.g. with respect to thinning, pruning and fertilizing.

Support for formation of cooperatives. It can be difficult for growers to develop forest owner cooperatives, and government can play an important role in facilitating this process.

Introduction of carbon trading. While obviously a hope for small-scale growers, as a means of private capture of a social externality, introduction of a viable carbon trading system presents many difficulties. If government can devise a system whereby small-scale growers can benefit from carbon credits, this may provide a major impetus to small-scale forestry.

Support for timber exporting. Policies which impede log exports, while well intentioned with respect to domestic employment and value adding, tend to maintain low timber prices and depress planting activity. Often an export market orientation is likely to stimulate a domestic timber industry. In this context, there are various measures which government can adopt to support new export initiatives (Ch. 9).

20.5 CONCLUDING COMMENTS

The foregoing discussion points to a promising future for small-scale forestry in the tropics. For many reasons, tree planting on farms can be expected to continue, and community forestry programmes are likely to expand, with considerable financial and environmental benefits. At the same time, a number of major challenges will be faced. However, improvements in tree technology, complementarity with other farming activities, the potential for growing trees on non-prime agricultural land, and the enthusiasm of participants, can be expected to lead to continued if not spectacular success in tropical small-scale forestry.

REFERENCES

Aggangan, R.T. (2001), 'Tree Farming in the Philippines: Some Issues and Recommendations', in S.R. Harrison and J.L. Herbohn (eds), *Socio-Economic Evaluation of the Potential for Australian Tree Species in the Philippines*, ACIAR Monograph No. 75, Canberra: ACIAR, pp. 33-43.

Dart, P.J., Brown, S.M., Simpson, J.A., Harrison, S.R. and Venn, T.J. (2001), 'Experience from ACIAR trials of the suitability and performance of Australian tree species in the Philippines', in S.R. Harrison and J.L. Herbohn, (eds), *Socio-Economic Evaluation of the Potential for Australian Tree Species in the Philippines*, ACIAR Monograph No. 75, Canberra: ACIAR, pp. 7-20.

Harrison S.R. and Herbohn, J.L. (eds) (2001), *Socio-Economic Evaluation of the Potential for Australian Tree Species in the Philippines*, ACIAR Monograph No. 75, Canberra: ACIAR.

Harrison, S.R., Herbohn, L.J. and Herbohn, K.F. (eds) (2000), *Sustainable Small-scale Forestry: Socio-economic Analysis and Policy*, Cheltenham: Edward Elgar.

Lamb, K. (2000), 'Implications for Small Scale Forestry of a Proposed Market for Trading Carbon Sequestration Credits from Forests in New South Wales Australia', paper presented to the IUFRO Small-scale Forestry Conference, Kuranda.

PCARRD-DOST-DENR-FMB-DA-CFNR-FDC/ENFOR (1999), *Guidelines for Watershed Management and Development in the Philippines*, Los Baños, Laguna.

Venn, T.J., Harrison, S.R. and Herbohn, J.L. (2001), 'Imperatives for reforestation in the Philippines', in S. Harrison and J. Herbohn, (eds), *Socio-Economic Evaluation of the Potential for Australian Tree Species in the Philippines*, ACIAR Monograph No. 75, Canberra: ACIAR, pp. 21-31.

Index